Books are to be returned on or before
the last date below.

18 NOV 1999

The centrifugal model testing
of waste-heap embankments

The centrifugal model testing of waste-heap embankments

YU. N. MALUSHITSKY

Edited by A. N. Schofield

Translated by David R. Crane

CAMBRIDGE UNIVERSITY PRESS
Cambridge
London   New York   New Rochelle
Melbourne   Sydney

Published by the Press Syndicate of the University of Cambridge
The Pitt Building, Trumpington Street, Cambridge CB2 1RP
32 East 57th Street, New York, NY 10022, USA
296 Beaconsfield Parade, Middle Park, Melbourne 3206, Australia

Russian edition © 'Budivelnik' Publishing House, Kiev, 1975
English translation © Cambridge University Press 1981

First published in Russian as *Ustojchivost' nasypej-otvalov (Tsentrobezhnoe modelirovanie)*, 1975
This translation first published 1981

Printed in Great Britain at the University Press, Cambridge
Typeset by H Charlesworth & Co Ltd, Huddersfield

Library of Congress cataloguing in publication data

Maliushifskiĭ, IUriĭ Nikolaevich.

The centrifugal model testing of waste-heap embankments.

Translation of Ustoĭchivost' nasypeĭ-otvalov.

Includes bibliography and index.

1. Soil banks–Models. 2. Soil stabilization.
I. Schofield, Andrew. II. Title.
TN292.M33313   628'.36   78-67431
ISBN 0 521 22423 3

Contents

Preface *page* ix

1 Model testing procedure 1
Basic principles 1
The centrifugal model testing installation of the Ukr. N.I.I. Projekt Institute 4
Types and dimensions of the models 8
Scales of modelling 11
Test procedures 17
The recording of observations 21
Documentation 27
The soil materials of the models 29
The allocation of operatives 37

2 The accuracy of the modelling of unconsolidated waste-heaps 40
The investigation of sloping models 40
The investigation of non-sloping models 48
Modelling the natural state 61

3 The relationship between the stability of unconsolidated waste-heaps and the moisture-content of the soils, their initial density in the waste-heap and the rate of formation of the waste-heap 72
Five basic factors determining the conditions of the stability of unconsolidated waste-heaps on a firm, horizontal foundation 72
Plan of the experiment 73
Determinations of the critical height H_{crit} and of the relationship $\alpha_{lim} = f(H)$ 78
Investigation of the influence of the moisture content and initial density of soils and of the rate of waste-heap formation on the stability of waste-heaps 84
The determination of $H_{crit} = f(w, \delta, Q)$ by the factor experiment method 93

Contents

4	**The influence of the composition of a soil mixture on the stability of an unconsolidated waste-heap**	99
	The soil mixture of the waste-heap	99
	The plan of the experiment	104
	The character of the variations in shear resistance of soil mixtures in relation to their composition	105
	The relationship between the critical heights of waste-heaps, the critical angles of the slopes and character of the failure of the waste-heaps on the one hand, and their soil composition on the other	109
	Conclusions from the tests with soil mixtures	113
5	**The scope for the use of the centrifugal model testing method for the determination of the influence on the stability of a waste-heap of different technological systems of waste-heap formation**	116
	Basic technological systems of waste-heap formation	116
	The plan of the experiment	117
	Results of the tests	120
6	**The stability of waste-heap embankments on weak, inclined bases**	125
	Basic principles	125
	Examples of the abbreviated determinations of the influence of z and β on the relationship $\alpha = f(H)$	126
7	**The conditions of the safe deposition of high waste-heap embankments onto weak (semi-liquid) foundations**	134
	Basic principles	134
	The resolution in models of the conditions for the safe deposition of 'dry' waste-heaps of non-saturable soils on old hydraulic waste-heaps	140
	Changes in the physico-mechanical properties of pulp soils when subjected to loads from waste-heaps	144
	Conclusions and recommendations	152
8	**The conditions of the stability of high waste-heaps on firm, inclined bases**	160
	The task and the programme of the experiment	160
	Results of the experiment	162
	Practical conclusions	170

Contents

9 Some recommendations regarding the use of the centrifugal model testing method in the solution of problems concerning the stability of waste-heap embankments   175
What may be, and what must not be resolved by the centrifugal model testing method   175
The plan and quality of an experiment   176
Starting materials for model waste-heap embankments   177
Concerning the prevention of certain errors which are possible when centrifugal model testing   177
Recommendations regarding the use of the present book as a handbook on method   180

Appendices   183
1. Example of the determination of the basic parameters of the slopes of model waste-heaps at the time of their failure   184
2. Example of the establishment of the degree of accuracy of the determination of the critical height of a waste-heap   186
3. Example of the verification of the significance of the variation in the average basic indices of the stability H_{crit} of sloping models of various height by the method of dispersion analysis   187
4. Example of the use of the factor experiment method to determine the influence of the rate of deposition, moisture content and initial density of the soil materials on the stability of unconsolidated waste-heaps in centrifugal model testing   189
5. The M.Ts.M.-2 dynamometer for the remote measurement of compression stresses in models (developed by G. E. Lazebnik)   198

Bibliography   200

Index   202

Preface

In the construction of hydrotechnical, transportation and mining structures the loss of stability by high embankments is accompanied in the majority of cases by collapses, all of which leads to enormous material expenditure and difficulties in carrying out restoration work. At the same time, industry is incurring multi-million rouble losses on account of its inability to take advantage of the surplus strength and the reserves of stability of embankments. Hence the establishment of the overall conditions of the stability of high embankments is one of the most important tasks in the construction and operation of earthen structures.

High embankments with a volume of many millions of cubic metres are continuously being built up during the construction and operation of open-cast mineral workings in the form of heaps of waste soil. The author has therefore made use of them as objects for a study of the conditions of stability of high unconsolidated embankments and for the development of effective means of improving those conditions and of reducing unproductive expenditure in the performance of dumping operations.

For example, merely by making use of the strength-recovery of soils under their own self-consolidation in the embankment, a saving which can be calculated in millions of roubles is achieved by reducing the volumes of material involved in the re-excavation of waste-heaps which are capable of maintaining stable slopes at angles steeper than the natural angle of repose assumed by soil when rolled down a slope. Or again, in another case, it becomes possible to prevent the irretrievable loss of thousands of hectares of useful land, buried beneath waste tips and embankments, which occurs as a result of inability to ensure stability and the safe performance of dumping operations on adjacent swamped dumping areas of old hydraulic waste-heaps.

The stability of high embankments at the time of their formation and during subsequent periods of the gradual consolidation of their body and foundation depends on many variable magnitudes, acting simultaneously and sometimes with opposing effects, which not only exert a direct influence on the stability of the embankment but also mutually affect each other. Hence the determination of the conditions of stability of an unconsolidated embankment proves to be a problem with many multi-stage unknown quantities, which do not lend themselves to calculation in analytical solutions.

From this arose the necessity to employ for the solution of such problems the method of centrifugal model testing, by means of which it is possible to solve three groups of problems concerning the stability of waste-heap embankments:

1. To determine the extent of the influence of concrete factors on the stability of waste-heap embankments.

2. To determine the degree of stability of waste-heap embankments in conditions of an actual open-cast mine or construction site.

3. To establish the indices of the relative stability of soils in waste-heap embankments so as to make use of them when planning waste-heaps as early as at the stage of the preliminary survey of the deposit sites or of a construction area.

This book represents the first systematised investigation carried out by the author of the stability of unconsolidated waste-heap embankments on both firm and weak, horizontal and inclined foundations by the centrifugal model testing method, and it can therefore be used as a textbook on method for the conduct of similar work. In it the method, developed by the author, is expounded and the limits of the accuracy of modelling are defined, methods are developed for determining the maximum guaranteed values of the critical heights of waste-heap embankments, given the variability of variable factors which determine the conditions of the stability of a waste-heap, and examples are given of the setting-up of an experiment and the processing of the associated calculations.

The practical significance of this work is to be found in the establishment of the necessary conditions for the prevention and elimination of landslips and the collapse of waste-heap embankments, and also in the establishment of the extent of the influence on stability of such factors as the moisture content of the soils, their initial density and lithologic composition, the nature of the foundation, and the rate and method of deposition of the embankments.

At the present time we can acknowledge with assurance the appropriateness of the use of the centrifugal model testing method for the solution of concrete problems concerning the stability of unconsolidated earthen and soil masses, for which there are no satisfactory general analytical solutions. It is necessary, however, to issue a caution about the erroneous notion concerning the centrifugal model testing method – the notion that it automatically provides prompt answers to any problems with the use of just one or two models, prepared from soils taken at random from the prototype waste-heap embankment.

The present book has been written on the basis of data collected as a result of many years of investigation conducted by the author using 255 models. As a result of these investigations by the author:

1. Measures have been worked out for the elimination of landslips at the waste-heaps of one of the open-cast natural sulphur mines, which have enabled the enterprise to save approximately 175 000 roubles per year, which were previously expended in combatting landslips.

2. Recommendations have been issued to reduce the volumes of waste material re-excavated from the internal dumps of the Krasnogor open-cast

Preface xi

coal mine (Kuzbass), and these yield an annual commercial saving to the value of 300 000 roubles.

3. A safe method of depositing high 'dry' waste-tips on abandoned areas of old hydraulic waste-heaps in the Kuzbass has been developed, which yields a saving on transportation of 250 000 roubles per year for a single hydraulic waste-heap and which frees large areas of useful land from the threat of being buried beneath new waste-heaps.

The author expresses his gratitude to: mechanic G. Kh. Pronko and Eng. A. P. Sakhno, who participated in the conduct of the experiments on models; Eng. A. A. Preobrazhensky, who assisted in the statistical processing of the results of the experimentation; V. S. Vagorovsky, Chief Engineer of the 'Krasnogor' open-cast coal mine, who brought about the implementation of our recommendations in industry; and especially to U.S.S.R. State Prize-winner, Honoured Scientist and Technologist of the R.S.F.S.R., Professor G. I. Pokrovsky, who has taken upon himself the labour of reviewing this book.

Kiev 1975
Yuri N. Malushitsky
Doctor of Technology

Preface to the English edition

The author expresses his sincere gratitude to Professor A. N. Schofield of the University of Cambridge who has played an active role in the publication of this monograph in an English translation, and also to the Cambridge University Press for assuming responsibility for this publication.

Kiev 1977
Yuri N. Malushitsky

1

Model testing procedure

Basic principles

In the mining industry, where waste soils are accumulated in high tips, attempts at analytical determinations of the stability of embankments in an unconsolidated state [1], [3], [4] are for the most part of theoretical interest only, since they are much too unwieldy for practical application and this fact, apart from the difficulties of the calculations, leads to an accumulation of hidden errors.

The simplest way of obtaining solutions that are acceptable for practical purposes is to use the model testing method, on condition that the models are constructed from the same soil materials that comprise the waste-heaps in the natural state. This condition is fulfilled by the centrifugal model testing method which was proposed in the thirties in three places at the same time: in the U.S.S.R. (Moscow, Professor G. I. Pokrovsky; Leningrad, Professor N. N. Davidenkov) and in the U.S.A. (Professor P. B. Bucky).*

As the criterion of approximate similarity between the natural object and the centrifuge model [13], [14], apart from geometrical similarity, the following conditions – sometimes referred to as the G. I. Pokrovsky criterion – have been adopted:

$$g_N \rho_N h_N = g_M \rho_M h_M$$

$$F_{0N} = F_{0M},$$

where g is the acceleration of the force of gravity or of the centrifugal force which replaces it in the model; ρ is the density of the soil; h is the vertical component of the deformation; F_0 is the specific free energy; N, M are signs indicating whether the relevant magnitude pertains to the 'natural' object or state, or to the 'model'.

The bulk forces of gravity which are acting in the natural state and object are replaced in the model by the vectorial sum of the force of gravity and the centrifugal force (given a vertical axis of rotation). The centrifugal force is normally many times greater than the gravitational force. In these conditions

---

*The original idea of reproducing in a model (made from an ideally elastic material) the action of the bulk forces of gravity was expressed in 1869 by the French academician E. Phillips [20]. He, however, did not extend his proposition to the field of the investigation of soils, and his idea remained unexploited until it was used in soil mechanics.

it can be reckoned approximately that the acceleration of the centrifugal force exceeds the acceleration of the earth's pull by as many times as the linear scale of the model is smaller than the scale of the natural object which is being modelled. The tangential accelerations, which arise during an increase or reduction in the speed of rotation of the centrifuge, are ignored.*

We propose to express the stability of an actual unconsolidated waste-heap in terms of the critical height of the slope H_{crit} (the critical height above which a slope, built up at the natural angle of repose of soil α_{rep}, begins to slump spontaneously) and by means of the relationship:

$$\alpha_{lim} = f(H),$$

where α_{lim} is the limiting resulting or general angle of the slope; H is the height of the slope, greater than H_{crit}.

The representativeness of the model of an unconsolidated waste-heap will obviously be determined by observance of the conditions of similarity

$$H_{crit\,N} = nH_{crit\,M}$$
$$\alpha_{lim} = f(H_N) = f(nH_M)$$

and by the condition of identity of the physical state and the properties of moisture content, density, and strength of the waste-heap materials at equivalent points (equivalent in terms of their position in time and space) of the natural object and the model

$$w_N, \delta_N, (\varphi,c)_N = w_M, \delta_M, (\varphi,c)_M,$$

where w is the moisture content of the soil, by weight; δ is the density, expressed in terms of the bulk weight of the skeleton of the soil; φ, c are the angular and linear indices of the shear resistance of the soil.

The magnitudes H_{crit} and the functions $\alpha_{lim} = f(H)$ are in direct proportion to those factors, the extent of whose influence on the stability of waste-heaps is one of the fundamental objectives of the investigations we are now describing. In the centrifugal model testing method this influence is determined by changing the quantitative value of one of the factors in a series of models under test whilst maintaining constancy of the magnitudes which characterise the remaining factors. These investigations can be expanded by finding an interconnection between the influences of two or several factors.

Until the present time only the qualitative aspects of an influence (plus or minus) were known, and then of only a few factors. Hence the establishment of even approximate laws governing the influence of the most important of these is essential for an understanding of the conditions of stability of unconsolidated embankments and can serve as a start in the search for quantitative evaluations

---

*For details of the principles of centrifugal model testing see the works of Professor G. I. Pokrovsky and Professor I. S. Fyodorov [13], [14], [15].

and for the limits of the influence of one or another factor, or of a group of factors, on stability.

The centrifugal model testing method has been in use for the investigation of the stability of dams and road embankments for more than forty years. Investigations of this kind have been, or are still carried out in the scientific-research institutes of VODGEO (Moscow), of Transport Engineering (Moscow), of Foundations and Underground Structures (Moscow) and in the scientific-research sectors of the Higher Educational Institutes for Transport: D.I.I.T. (Dnepropetrovsk) and M.I.I.T. (Moscow), and also in the Hydroprojekt Institute named after S. Ya. Zhuk (Moscow). However, neither in our own native practical experience, nor abroad, are any analogies known for the investigation by the centrifugal model testing method of the stability of unconsolidated soil waste-heaps, with the exception of the unique case of the determination, in 1956-7 by V. P. Zapolsky (with our participation) in the Krivoi Rog Scientific-Research Institute for Mineral Ores, of the optimum angles of the slopes of waste-heaps which were included in the plan of a group of open-cast manganese mines in the Nikopolsk deposits.

The investigators are making use of the model testing system proposed by Professors G. I. Pokrovsky and I. S. Fyodorov [13], [14], [15]. We shall now introduce a brief summary of the starting principles adopted by the authors of the method.

The materials of the model and the natural object are identical.

The modelling of soil slopes can be only approximate, since given the identical nature of the materials of the natural object and the model it is impossible to observe all the conditions of the total criterion of similarity.

As the criterion of the approximate similarity between the natural object and the model we are adopting G. I. Pokrovsky's criterion.

We accept the principle established by experience that, given a sufficient width of the model, the friction between the model and the walls of the container (packages) does not exercise any noticeable influence on the deformations of a slope when model testing, and that it can therefore be ignored.

When modelling an artificially deposited embankment one should as far as possible reproduce the method of deposition of the soil. Moreover, by increasing the speed of rotation of the model and with a corresponding increase in the scale of the structure (of its effective height) the rate of deposition of the structure is also modelled.

The problem of finding the critically steep angle of a soil slope is resolved by bringing the slope to the point of collapse by gradually increasing the centrifugal force.

The linear scale of modelling n is determined with sufficient accuracy for practical purposes in accordance with the formula

$$n = \frac{N^2 R}{894},$$

where N is the number of revolutions of the model per minute; R is the effective radius of a given point of the model, equal to the distance between that point in a radial section of the model and the axis of rotation, in metres.

In general cases the scale of time τ for the centrifugal modelling of soil waste-heaps varies. Depending on the soil composition and its physical state, τ can vary within the limits n^0-n^2.

The method followed in our investigations is based on the principles set out above and which have been approved by the experience of our predecessors. Deviations from them have been permitted only in those cases when discrepancies between one or another principle and the factual data have been discovered or when we have succeeded in introducing into them certain refining limitations or additions. The experiments described were carried out at the centrifugal installation of the Ukr. N.I.I. Projekt Institute in Kiev.

The centrifugal model testing installation of the Ukr. N.I.I. Projekt Institute

In order to resolve practical problems concerning the stability of soil waste-heap embankments, under the direction of the author (Yu. N. Malushitsky), and with the participation of Giprostrommash Institute and the 'Leninskaya Kuznitsa' factory, a unique centrifugal installation was built in the Institute of Ukr. N.I.I. Projekt and brought into operation in 1966.

The centrifuge installation occupies several specialised rooms: for the preparation of the materials for the models, for storing these materials in regulated moisture conditions, for the preparation and measurement of the models, a mechanical workshop, a tensometric room, rooms for analytical and soil survey investigations and a separate machine hall (fig. 1), at the centre of which in an underground chamber the centrifugal machine is installed (fig. 2).

The machine (fig. 3) consists of a drive, reduction gear, and a vertical shaft with a balanced arm and carriages suspended from it, which carry the container with the models.

The electrical drive for the machine is achieved by means of a combined generator–motor system. A 115 kW D.C. motor is located in close proximity to the centrifuge. The generator-motor system is arranged to one side in the machinery hall (see fig. 1). The control desk is sited in a separate room adjacent to the machinery hall and linked to it by two inspection windows.

The loading of the heavy containers with the models is effected via a hatch in the cover of the centrifuge chamber with the aid of a 3-tonnes electric overhead hoist and horizontal thrust pulleys.

Model testing procedure

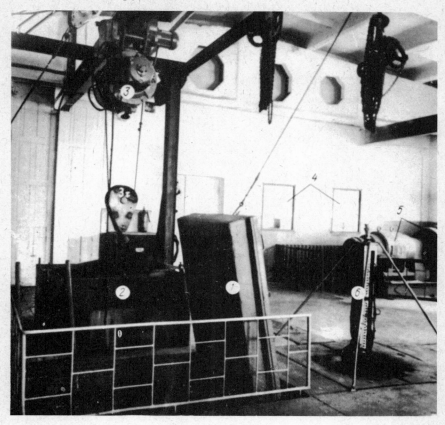

Fig. 1. The machine hall of the centrifugal installation: 1, the hatch cover from the entrance into the test chamber; 2, the container for the model; 3, overhead lifting gear with a lifting capacity of 3 tonnes; 4, windows from the control desk; 5, the generator-motor system; 6, electrical slip rings.

The technical and operational characteristics of the centrifugal installation

1. The maximum scale of modelling on this assembly is 1 : 320.
2. The maximum (rated) number of revolutions per minute of the centrifuge is 340.
3. The effective radius of the machine is 2505 mm.*
4. The maximum internal dimensions of the containers for the models (without any lining) are: length 1400 mm, width 500 mm, height 750 mm.

*The distance from the centre of the mass of a fully loaded container (for models) to the axis of rotation of the model.

Fig. 2. The underground chamber with the centrifugal machine: 1, the axial shaft; 2, the balanced arm; 3, the pivot journal; 4, the carriage with container.

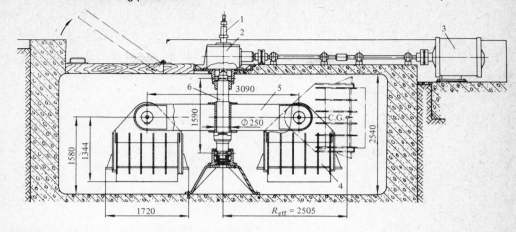

Fig. 3. The centrifugal machine of Ukr. N.I.I. Projekt: 1, electrical slip rings; 2, reduction gear; 3, drive motor; 4, carriage with container; 5, balanced arm; 6, shaft. Dimensions in millimetres.

5. The rated bulk weight of the material comprising the models (when the whole volume of the container is filled) is 2.0 tonnes/m^3.

6. The weight of the centrifugal machine itself (without models) is 18.7 tonnes.

7. The axial shaft of the centrifuge is securely mounted in bearings at its upper and lower ends.

8. The dimensions of the underground chamber are: diameter 6700 mm, height 2540 mm. The axis of the chamber coincides with the axis of rotation of the machine.

9. The electrical drive of the machine, consisting of a generator-motor system, permits smooth control of start-up and braking and is arranged in cascade: (*a*) a 160 kW capacity electric motor, operated from a 380 volt A.C. power supply at 970 revs/min; (*b*) a 135 kW capacity generator, operated from a 220 volt D.C. power supply at 1000 revs/min; (*c*) a 115 kW capacity electric motor, operated from a 220 volt power supply at 870 revs/min.

10. By means of channels in the axial shaft of the machine and in the slip-ring column, electrical contact is maintained between the control desk and the models situated in the centrifuge carriages throughout the whole time they are under test, to permit remote observation of the state of the models; hydraulic contact is similarly maintained to permit the supply of water to the models or the additional loading of the models with shot.

11. Remote observation of the operation of the machine is achieved with the aid of electrical instruments at the control desk. However, these can be duplicated by visual observations through the window in the inspection well; in the latter case a stroboscope must be installed to provide synchronised illumination of the rotating model by means of light flashes.

Types and dimensions of the models

Starting from the magnitude of the calculated effective radius of the centrifugal machine, equal to 2.505 m, and the limitations of five percent recommended by the authors of the centrifugal modelling method [14] on the permissible deviations from the average in the accelerations of the boundary points of the model, the maximum height of the model was determined as 35-36 cm. However, the fundamental experimentation was carried out using models of 24 cm in height. This reduced the deviation from the average to 3.5-4.0%.

In our investigations the maximum permissible width of the models allowed by the design of the centrifugal installation was taken as 48-49 cm, based on the following experiment.

The first three model waste-heaps (nos. 1, 2, 3) of triangular cross-sectional shape (fig. 4) and each with a height of 18 cm were made in three different widths (in two containers): 48.3 cm (in one container), and 36.0 cm with 11.5 cm (in the other container). The models, which were constructed from precisely the same soil material, were thus centrifuged simultaneously.

The application of load was carried out in five successive runs on the machine, each lasting a time of $t = 19'12''$ and achieving a maximum number of revolutions per minute of 135 in the first run, 190 in the second, 232 in the third, 269 in the fourth and 300 in the fifth, which corresponded (where $R_{eff} = 246.5$ cm) to scales of 50, 100, 150, 200 and 250.

After each run the models were measured and topped up to their original profile. The width of model no. 1 (48.3 cm) permitted measurements to be taken on two sections (A and B). Models nos 2 and 3 were measured once only – on a section passing through the middle of each model.

The results of the tests, set out in table 1, showed the partial suspension of

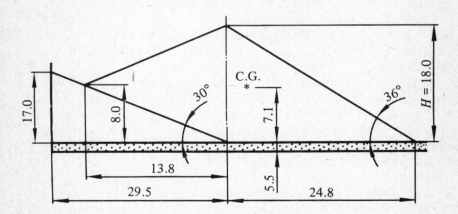

Fig. 4. Cross-sectional profile of a model waste-heap of triangular section. Dimensions in centimetres.

Table 1 Results of the tests on models nos. 1, 2 and 3

Loadings (runs)	Maximum number of revolutions per minute	Scale of modelling	Model numbers												
			1A	1B	2	3	1A	1B	2	3	1A	1B	2	3
			Height H_N (metres)				Settlement (metres)				Angles of slope, α_{av}			
I	135	50	6.25	6.75	6.05	7.50	2.75	3.25	2.95	1.50	28°00'	27°10'	26°45'	27°55'
II	190	100	13.2	13.2	13.2	13.7	4.8	4.8	4.8	4.3	27°50'	28°40'	28°20'	29°15'
III	232	150	19.8	19.7	20.4	21.0	7.2	7.3	6.6	6.0	28°50'	27°30'	29°35'	29°45'
IV	269	200	27.0	26.6	27.6	28.6	9.0	9.4	8.4	7.4	26°55'	28°20'	26°15'	29°20'
V	300	250	36.3	34.8	a	a	8.7	10.2	a	a	25°50'	26°10'	a	a

[a] On run V the partition wall collapsed in the container carrying models nos. 2 and 3.

Note: Initial data: $H_M = 18$ cm, $w = 26.8\%$; loose fill; duration of the runs $t = 19'12''$; Kerchi clay; width of the models, in cm: nos. 1A and 1B, 48.3; no. 2, 36; no. 3, 11.5.

the model with a width of 11.5 cm between the side walls of the container and, as a result of this, a rate of settlement of the model which lagged by an average of 7.3% in comparison with the settlement of the model with a width of 48.3 cm.

The lag in the settlement of the model with a width of 36.0 cm in comparison with the maximum deformations of model no. 1 was significantly smaller, but during the first two loadings (runs) the deformations of model no. 2 did not exceed the limits of the average deformations of model no. 1.

The partial suspension of the model between the side walls of the container imparts to the deforming model the shape, in plan, of a protruding, sliding tongue, with minimum extension in the region of the side walls. In the middle third of the models of great width the tongue of slipped earth all extends the same distance.

A horizontal cross-section of the model at the level of half its height (fig. 5) reveals a surface of slip with distinctly delineated curves, in plan, over a distance of 8–12 cm from the walls of the container, brought about by the retarding action of friction. Between these marginal sections the displacement remains the same and consequently corresponds to the conditions of a plane model.

As a result of this experiment and of subsequent verifications with cross-sections of models it was decided to conduct the tests on models using the full width of the containers: for the first 48.8 cm, for the second 49.2 cm, taking only the central part of the models as free from the distorting influence of the side walls and carrying out the measurements on the cross-sectional sections of all the models through the lines A and B, which were at a distance of 16 cm from the two side walls.

Transparent plastic was adopted for the lining of the walls of the container, without a lubricant in the case of sandy-clay soils on account of the undesirability of introducing any additional medium (graphite or oil) into the soil mass of the waste-heap undergoing test.

For the purposes of comparison of the results obtained in determining H_{crit} and $\alpha_{lim} = f(H)$ on comparable models the constant form of the original section of a sloping model was adopted, as this satisfies the conditions of the preservation of the geometrical similarity of the cross-sectional areas throughout the whole

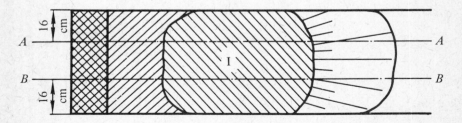

Fig. 5. Horizontal cross-section of a model waste-heap after collapse (slipping). I is the collapsing body of soil.

Model testing procedure

period of model testing, right up to the onset of the collapse of the model.

The models were also strictly controlled with regard to the constancy of the composition of the soil, its moisture content, the degree of its pulverisation and the density of its initial placement into the model, and also with regard to the method and speed with which loading was applied (acceleration of the machine).

When resolving practical problems it is appropriate to ensure that all these conditions are as close as possible to the average conditions of prospective natural waste-heaps. As typifying such an average we adopted a block waste-heap of height H, deposited by using waste-heap forming machinery with a productivity rating of 800–6000 m³/hour, with a horizontal surface at the level of the upper brow, with an angle of slope equal to the natural angle of repose of soil α_{rep} plus one degree, and with an over-spill width of newly covered land $B \approx 1.5\,H$.

When the waste-heap is being dumped, part of the soil falls on the slope of the waste-heap formed by the previous overspill. Therefore in order to exclude the effect of the scale factor when determining the critical height of the model it is necessary to ensure that, when the centrifugal machine is set in motion, the position of the brow of the slope should not change in relation to the point (line) where the horizontal foundation intersects the plane of the slope of the waste-heap formed by the previous overspill.

Since in the natural state, as a result of self-consolidation, the slope of the waste-heap formed by the previous overspill slumps somewhat by comparison with the angle of repose, the slope of the waste-heap formed by the previous overspill in the model was given an angle 5–6° less than the natural angle of repose.

In our tests the angles of repose were determined as equal to 35°, which corresponded to the measurements taken in the natural state of waste-heaps of Novy Razdol quaternary and tertiary soils and Kerchi tertiary clays. The waste-heap formed by previous overspill is conventionally considered as fully consolidated and incompressible. In the model this slope is simulated by a surface of wooden blocks inclined at an angle of 30° and placed edgewise (fig. 6).

For the investigation of the changes in the physico-mechanical properties of the waste-heap soils within the depth of the body of the waste-heap as its height increases, a type of slopeless model was adopted having the form of a rectangular parallelepiped identical with the sloping model in height, material, moisture content, initial density of deposition of the soil, and system of load application.

Scales of modelling

The centrifugal model testing installation of Ukr. N.I.I. Projekt permits the speed of rotation of the models to be increased to 340 revs/min, but the conditions governing the accuracy of the readings of the tachometer used require a reduction in speed to 300–320 revs/min. Given an effective radius of the

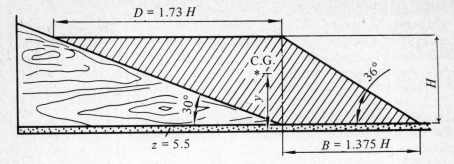

Fig. 6. Cross-sectional profile of the basic form of the models of waste-heaps of clayey soils.

machine R_{eff} = 250.5 cm, the maximum scales of acceleration can be increased to $n \approx 300$.

Given a height of the model H_M, the accelerations in the centre of gravity of a section of the model and at its extreme upper and lower points are different during rotation, and accordingly the bulk forces acting in them and the scales of modelling for them are also different.

If, in the centre of gravity of a section of the model, the scale of modelling is

$$n = \frac{N^2 R}{894},$$

then, in as much as the centres of gravity in the basic types of our models are to be found close to the middle of their height, the scales of modelling for the boundary points can be found according to G. I. Pokrovsky [14].

For the base of the model

$$n_{low} = n + n \frac{H_M}{2R}.$$

and for the upper part of the model

$$n_{upp} = n - n \frac{H_M}{2R}.$$

The magnitude of the distortion of the scale in the boundary points by comparison with the scale for the centre of gravity is equal to

$$\Delta n = \pm \frac{H_M}{2R}.$$

In the installation of the Institute Ukr. N.I.I. Projekt the height of the model which corresponds to the normally [14] permissible distortion of the order of 5% is

$$H_M = \frac{5 \times 2R}{100} = 0.1 \times 250.5 \approx 25 \text{ cm}.$$

Model testing procedure

In our investigations a height of 24 cm was adopted for the basic models, and only during comparative experiments on the 'modelling of models' were heights up to 36 cm permitted, these having deviations in the scales of modelling of up to 7% at the boundary points.

The scale of time τ can be expressed in terms of the scale (accelerations) of modelling n as:

$$\tau = \frac{T_N}{t_M} = n^x,$$

where T_N is the time for the occurrence of the observed process in the natural state; t_M is the time taken for the same process in the model; x is an index of degree, dependent on the material of the natural object and the model.

In the work cited [14] the relevant considerations concerning the scale of time when model testing are summarised in the following manner.

If a model, reduced in size by n times in comparison with the natural object, is subjected to stresses, created by the acceleration of the centrifugal field, equal to $a = ng$, where g is the acceleration of the force of gravity, and if geometrical similarity of all the parts of the system being modelled is preserved, then the time necessary for a mechanical displacement in the model is also equal [13] to $t_M = T_N/n$. The latter follows from Newton's second principle of mechanics.

R. P. Kaplunov and I. M. Panin [5] have proposed a method, within the limits of the strength of soil, of determining the scale of time by observations of the deformations of the natural object and the model. Their arguments can be summed up as follows.

The deformation λ as a function of the load (or stress) σ and of time t is expressed as

$$\lambda = \kappa_1 \sigma + \kappa_2 \sigma t,$$

where $\kappa_1 \sigma$ is the immediate deformation following a suddenly applied load (κ_1 is a constant magnitude, the inverse of the modulus of elasticity of the material); $\kappa_2 \sigma t$ is the deformation dependent on time (κ_2 is a non-constant magnitude which varies with time).

If the deformations of the natural object and of the model are

$$\lambda_N = F(T_N),$$
$$\lambda_M = f(t_M),$$

then they can be represented graphically.

Given the condition of similarity, for any material $\lambda_N = n\lambda_M$, from which

$$F(T_N) = nf(t_M).$$

The latter expression is illustrated by a graph (fig. 7) and by means of this we can determine the scale of time $\tau = T_N/t_M$.

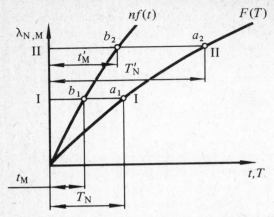

Fig. 7. The cited graph of the deformations of the natural object and the model in time.

Given the identical and constant stressing of the material of the natural object and the model, τ ought to be a constant magnitude,

$$\tau = \frac{T_N}{t_M} = \frac{T'_N}{t'_M} = \frac{T''_N}{t''_M} = \cdots = \frac{T_N^n}{t_M^n}.$$

Where n is large the values T_N and t_M can differ greatly. Hence their graphical representation becomes more difficult. In such cases along the axis of the abscissae one can plot not t, but log t.

In practice the latter equation cannot always be observed on account of unavoidable errors in experimentation. In that case Prof. R. P. Kaplunov suggests that the value of the scale of time should be determined in accordance with the laws of mathematical statistics, as the general average τ_{av} of the aggregate series τ_i.

Where x = const. and at any value of n, the value of τ as the scale of time will be valid for any values of T_N and t_M, and accordingly for the moment of the collapse of the model also.

Despite all temptation the Kaplunov-Panin method is inadmissible for the determination of the scale of time in our case of an unconsolidated waste-heap, in accordance with the limiting conditions imposed by the authors of the method.

The transition from one scale of modelling to another during the process of the model testing of an unconsolidated waste-heap occurs gradually and at different speeds at different levels of a section of the model. Whilst the soil mass in the upper, less heavily laden layers preserves its three-phase state throughout the whole model testing process, in the lower layers, with the increase in the scale of the accelerations, the consolidation of the soil may be attended by a transition to a two-phase state with a gradual transposition upwards from below, of the boundary between the two-phase and three-phase states of the soil material in the body of the model.

Model testing procedure

Accordingly, different scales of time can correspond to different parts of a model at one and the same time.

It would be expedient to try to arrive at some intermediate scale which would correspond, at any given moment, to the correlation between the parts of the model which happen to be in different states. It is obvious, however, that it is impossible to make such a calculation during the process of centrifugal model testing and consequently it is impossible to make any alterations in the control of the scale of time, all the more so since with the commencement of the collapse of the model, with the appearance in it of the first cracks, the reverse process begins – the transition of the body of the model from a two-phase to a three-phase state.

Prof. I. S. Fyodorov recommends, on the basis of many years of experience of the centrifugal model testing of earthen structures in the Vodgeo Institute, that in the case of models of embankments a constant scale of time equal to the first power of the scale of accelerations, should be maintained throughout the whole period of the model testing. We too arrived at the same conclusion as a result of the first stage of our investigations, with degrees of water saturation up to $w = 0.75$.

Convincing proof of the correctness of that conclusion was the fact that models of different initial heights, modelled by us in systems determined by the scales where $\tau = n$, gave identical results when converted to the natural state. Tests involving the 'model testing of the models' enabled us to carry out this check.

The comparative data relating to the moment of the commencement of collapse of twelve identical models constructed in three different scales are set out in table 2, whilst the time of the model testing can be determined by reference to the composite graph of the increase in accelerations of the centrifuge, depicted in fig. 8.

The indices of the first group of models (nos. 8, 9, 12, 13) at the moment of their collapse were: the average scale of modelling $n^I = 186.9$; average height of the models $H_M^I = 21.1$ cm; average duration of the increase in accelerations (of modelling) $t^I = 25.6$ min. The indices for the second group of models (nos. 10, 11, 14, 15) were correspondingly: $n^{II} = 148.9$, $H_M^{II} = 25.9$ cm; $t^{II} = 32$ min. And for the third group: $n^{III} = 124.3$; $H_M^{III} = 31.2$ cm; $t^{III} = 38.6$ min.

Taking the scale of time as $\tau = n$ minutes, we can find the corresponding magnitudes for the 'natural object':

$H_N^I = 186.9 \times 21.1 = 3921$ cm; $T_N^I = 186.9 \times 25.6 = 4785$ min;

$H_N^{II} = 148.9 \times 25.9 = 3873$ cm; $T_N^{II} = 148.9 \times 32.0 = 4765$ min;

$H_N^{III} = 124.3 \times 31.2 = 3896$ cm; $T_N^{III} = 124.3 \times 38.6 = 4790$ min.

Thus the duration of the loading of the waste-heap required to bring about the collapse of the 'natural object' T_N remains constant in all three cases.

Table 2 The scale of modelling and the heights of sloped models, at the moment of commencement of their collapse

Indices	Model no.				Average value of the indices	Model no.				Average value of the indices	Model no.				Average value of the indices
	8	9	12	13		10	11	14	15		6	7	16	17	
Height of the models, in cm:															
(a) initially	24.0	24.0	24.0	24.0	24.0	30.0	30.0	30.0	30.0	30.0	36.0	36.0	36.0	36.0	36.0
(b) at the commencement of collapse[a]	$\frac{21.1}{21.0}$	$\frac{21.3}{21.0}$	$\frac{20.9}{20.7}$	$\frac{20.7}{-}$	21.1	$\frac{26.5}{26.7}$ 24.7	$\frac{25.4}{-}$	$\frac{25.7}{27.6}$	$\frac{25.4}{-}$	25.9	$\frac{30.2}{31.0}$	$\frac{30.6}{32.1}$ 32.0	$\frac{31.0}{32.5}$	?	31.2
The scale of modelling at the commencement of collapse	$\frac{187}{187}$	$\frac{187}{187}$	$\frac{188}{184.5}$	$\frac{188}{-}$	186.9	$\frac{151.5}{154}$ 151.5	$\frac{150}{-}$	$\frac{151.5}{131}$	$\frac{151.5}{-}$	148.9	$\frac{126.8}{126.0}$	$\frac{126.8}{126.0}$ 119.5	$\frac{126.1}{118.5}$	?	124.3

[a]The numerator denotes the first signs of movement, the denominator indicates the development of displacement.

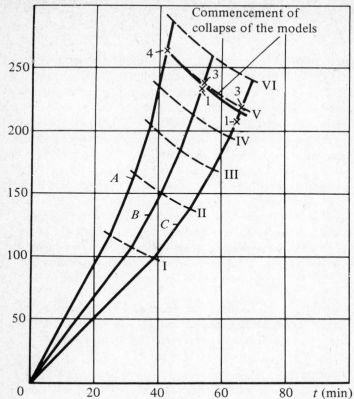

Fig. 8. Composite graph of the increase in accelerations of the centrifuge ($\tau = n$) for models nos. 6–17.
× = the recorded points of the commencement of collapse of the models and their numerical value; A, models nos. 8, 9, 12, 13 ($H_{init\ M} = 24$ cm); B, models nos. 10, 11, 14, 15 ($H_{init\ M} = 30$ cm); C, models nos. 6, 7, 16, 17 ($H_{init\ M} = 36$ cm); I, II, III, IV, V, VI = ends of first, second, third, fourth, fifth, sixth runs.

Consequently the adopted scale of time $\tau = n$ minutes for the model testing of the waste-heaps can be considered acceptable.

Test procedures

The rate of increase of the bulk loadings which occurs in the natural state during the formation of a waste-heap is modelled in the centrifuge by the rate of increase in the acceleration of the model waste-heap, and the height of the waste-heap at any given moment is defined by the speed of rotation of the model.

The average duration of the loading or acceleration of the model from 0 to the nth scale, given the scale of time $\tau = n$ established above, is defined as

$$t_M = \frac{2T_N}{n}.$$

In cases when soils with a high initial degree of consolidation are deposited in an embankment, full loading of the models to the point of collapse or up to a predetermined scale, whilst preserving an angle of slope close to the magnitude of the angle of repose, can be accomplished in one single, continuous acceleration of the models. However, when models are formed of loosely heaped material they can undergo intensive consolidation during centrifuging, with a considerable flattening of the angle of slope in their upper parts, whilst at the same time the angle of slope in their lower portions remains equal to the angle of repose. In such a case a model should accordingly not be loaded to the calculated height in a single run on the centrifuge.

In the natural state the formation of a waste-heap occurs with the simultaneous consolidation of the lower levels and an increase in the height of the waste-heap. In order to approximate to the conditions of the increase in height in both the model and the natural state it is necessary to stop the centrifugal machine several times and to add fresh soil to the model to top it up to its original profile.

If during the first run, or acceleration, the model achieves the scale n, then in repeated runs the scales of modelling are increased successively to $2n$, $3n$, $4n$, $\cdots$, mn. The duration of the runs, however, remains constant, since

$$t_{M-m} = \frac{2T_N m}{nm} = \frac{2T_N}{n}.$$

During the repeat runs the height of the lower portion of the slope, which exhibits an incline equal or close to the angle of repose, gradually increases until collapse of the slope begins at an angle whose value is $\alpha \approx \alpha_{rep}$.

In order to achieve the correct initial cross-sectional profiles the models are formed and topped up with fresh material with the use of the appropriate templates.

If during a scheduled stopping of the centrifuge any bulging of the slope in the model is discovered, beyond the limits of the original contour, then the supplementary soil is added with the aid of a specially trimmed template or mould, which restores the originally established brow and slope of the upper part of the inclined face of the model up to the point of intersection with the protruding part of the slope.

The cycle of loading of a waste-heap (fig. 9) formed by an overspill of width $CD = C'D'$ moving in the direction from the section $ABDC$ to the section $A'B'D'C'$, is completed at the moment of displacement of the vertical projection of a point on the brow of the waste-heap B (B') from point C to point C'.

The volume of a waste-heap (V) of height H, which corresponds to a full

Model testing procedure

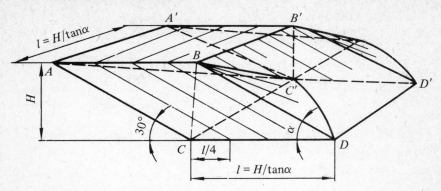

Fig. 9. Determination of the cycle of the loading of a waste-heap.

cycle of loading, given the types of waste-heaps adopted (in accordance with fig. 6), is defined as

$$V = \frac{(1.73 + 1.375) H^3}{2 \tan 36°} = 2.15 H^3,$$

whilst the length of time taken in the deposition of this volume (T), given a waste-heap formation capacity (Q), is equal to

$$T = \frac{2.15 H^3}{Q}.$$

In our investigations of the accuracy of modelling we adopted as the constant rate of waste-heap formation $Q = 3140$ m³/hour, whilst the initial heights of the models and the average scales of modelling varied.

Given the initial heights: $H_M^I = 24$ cm, $H_M^{II} = 30$ cm and $H_M^{III} = 36$ cm, and average scales of modelling of $n_I = 150$, $n_{II} = 120$ and $n_{III} = 100$, the time of loading or accelerating the models t_M is determined from the calculation:

$$t_M = \frac{2 \times 2.15 H_N^3}{Qn} = \frac{4.3 H_M^3 n^2}{Q}.$$

Equal to

$t_M^I = 0.427$ hours $= 25'36''$,

$t_M^{II} = 0.533$ hours $= 32'00''$,

$t_M^{III} = 0.640$ hours $= 38'24''$.

In the basic experiments, on the other hand, the initial height of the models remained constant at $H_M = 24$ cm, whilst the productive capacity of the nominal waste-heap forming machinery varied:

Q_I = 800 m³/hour of loose material,

Q_{II} = 2400 m³/hour of loose material,

Q_{III} = 5950 m³/hour of loose material.

The latter magnitude (Q_{III}) was determined by the conditions of the minimum duration of a run on the centrifugal machine necessary to achieve maximum revolutions, equal to 6 minutes.

For these conditions the duration of the runs on the centrifuge were determined in accordance with the foregoing:

$t_M^I = 44'48''$,

$t_M^{II} = 14'48''$,

$t_M^{III} = 6'00''$.

In tables 3 and 4 the preselected schedules for the loading or acceleration of the models in groups I and II are set out. When the acceleration of a model involves periods (level stages in the acceleration curve), during which a constant number of revolutions is maintained, the duration of this period is calculated for the natural state in accordance with the formula $T_N = nt_M$.

In order to determine the influence of the methods of waste-heap formation on the stability of a waste-heap the methods of depositing the heaps were

Table 3 *Calculated loading schedules (runs) for models of group I*

Loadings (runs)	Maximum number N (revs/min)	Limit scale of modelling n	Equivalent height of the waste-heap H_N (metres)	Maximum number N (revs/min)	Limit scale of modelling n	Equivalent height of the waste-heap H_N (metres)	Maximum number N (revs/min)	Limit scale of modelling n	Equivalent height of the waste-heap H_N (metres)
	Models nos. 8, 9, 12, 13			Models nos. 10, 11, 14, 16			Models nos. 6, 7, 16, 17		
I	117	37.5	9	106	30	9	98	25	9
II	167	75	18	150	60	18	138	50	18
III	204	112.5	27	184	90	27	169	75	27
IV	235	150	36	212	120	36	195	100	36
V	263	187.5	45	237	150	45	218	125	45
VI	288	225	54	260	180	54	239	150	54
VII	311	262.5	63	281	210	63	258	175	63
VIII							276	200	72

Note: The indices for models nos. 8, 9, 12, 13 were $H_{init\ M}$ = 24 cm; R_{eff} = 2.41 m; duration of each run t = 25'36''.

Those for nos. 10, 11, 14, 16: $H_{init\ M}$ = 30 cm; R_{eff} = 2.38 m; duration of each run t = 32'00''.

Those for nos. 6, 7, 16, 17: $H_{init\ M}$ = 36 cm; R_{eff} = 2.35 cm; duration of each run t = 38'24''.

Table 4 Calculated loading schedules (runs) for models of group II (basic) with an initial height $H_{init\ M}$ = 24 cm and effective radius R_{eff} = 2.41 m

Loadings (runs)	Maximum number N (revs/min)	Limit scale of modelling, n	Equivalent height of the waste-heap, H_N (metres)	Duration of each loading (run) t (min and sec), at a predetermined rate of deposition Q (m³/h)		
				Q_I = 800	Q_{II} = 2400	Q_{III} = 5950
I	96	25	6.0	t_I = 44'48"	t_{II} = 14'48"	t_{III} = 6'00"
II	136	50	12.0			
III	166	75	18.0			
IV	192	100	24.0			
V	215	125	30.0			
VI	236	150	36.0			
VII	255	175	42.0			
VIII	273	200	48.0			

modelled with successive changes of the templates or moulds for the cross-sectional sections of the models during their initial deposition and during subsequent topping up. In these cases the loading procedures were selected to correspond to the conditions in the natural state.

The recording of observations

In the investigations conducted the recording of observations was carried out in three ways: by taking static measurements of the profiles of the models before and after tests; by the remote measurement of vertical and horizontal deformations of the models whilst under load (during a run) and by recording the physico-mechanical properties and condition of the soil materials of the models before and after the tests.

Measurements of the profiles of the models before the commencement of tests and after each loading (each run on the centrifuge) were taken with the aid of the automatic profilometer (fig. 10).

The profilometer takes the form of an assembly of 63 light metal rods with rubber tips, arranged at 2 cm intervals in a vertical plane and held between two blades with the aid of compression springs. The springs are disengaged by means of a lever, located on one of the compression blades, and the rods are then able to move freely between the blades.*

In order to take a cross-sectional profile the profilometer is placed above the model with its ends resting on the face walls of the container. The rods are then released from the pressure of the blades and under their own weight they drop down until checked by the surface of the model; they are then again clamped in this position. The profilometer is then removed and placed on a sheet of

*The rubber tips serve to prevent the rods from penetrating into the soil.

Fig. 10. Plotting the profiles of models with the aid of the automatic profilometer.

millimetre-squared graph paper and the cross-sectional profile of the model is then sketched in by following the ends of the rods in their fixed position. By superimposing the cross-sectional profiles recorded before testing and after each loading we obtain a picture of the deformations which the model undergoes throughout the whole period of the tests. An example of superimposed profiles is shown in fig. 11.

When the tests are completed sectional cuts are made through the model in the vertical and horizontal planes. In these planes contact prints or sketches are made of the position of the boundaries of the surfaces of slip in cross-section and in plan; the latter is made at a height of $0.416\ H_{init}$ from the base of the model.*

The vertical and horizontal deformations of the model during the period of increase of the bulk forces when centrifuging are measured remotely with the aid of electrical settlement indicators (S.I.) and displacement indicators (D.I.).

The information from the indicators is conveyed to the measurement panel via an electrical circuit whose continuity at the point of transition from the rotating model to the stationary panel is achieved by means of a multi-channel current collector. The latter consists of an assembly of cupro-graphite brushes which slide over bronze rings which are rotating together with the model and to which the electrical leads from the individual indicators are connected.

*Where the initial height of the model is 24 cm the horizontal section is made at a height of 10 cm from the base.

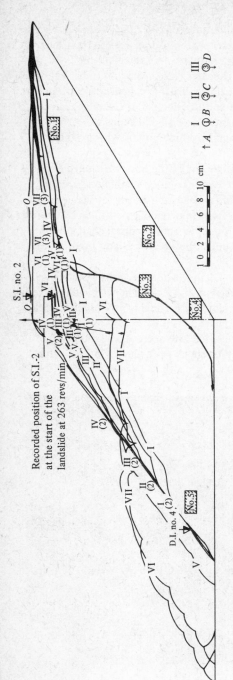

Fig. 11. Superimposed profiles of the measurements of model no. 13.
O, initial contour of the model; I, the contour after run I; II, the same after run II; III, after run III; IV, V, VI, VII respectively. A = datum mark, fixing the position of the initial brow of the slope; B = points where measurements of the height of the slope were taken; C = measuring points for the height of the terrace formed by bulging; D = measuring points at the cleft; nos. 1–5 = test sampling points.

Centrifugal model testing of waste-heap embankments 24

The first indicators for recording the horizontal displacements of the models were based on the principle of a commutator switch sliding over contact buttons and linked to small control light bulbs. Subsequently we rejected these inadequate indicators and went over to a rheostat type which in the beginning was used only to record settlements. Rheostat indicators permit the continuous recording of observations. They are based on the principle of the change in the voltage of a direct current through the constantan coil of a rheostat.

The designs of the S.I. and D.I. indicators developed by G. Kh. Pronko are shown in fig. 12.

In order to reduce the effect of the self-weight of the indicators on the model the moving parts of the indicators are fitted with counterweights balanced in such a way that the weight of the indicators exceeds the weight of the counterweights by a mere 5 g. The S.I. indicators are supported on the model by a wide wheel which rolls over the surface of the model when horizontal displacements of the model occur. Both types of indicator can be located at any place on the model.

The readings from the indicators are registered, with a record of time lapse, on automatic recording instruments of the type N-370-M. The wiring circuit and the method of switching the indicators devised by Engineer G. N. Bychkovsky is shown in fig. 13.

The indicators are calibrated when connected into the overall system of remote measurement. Hence the necessity to check the accuracy of individual elements within the system does not arise.

Fig. 12. Installation of S.I. and D.I. indicators in models.

Model testing procedure

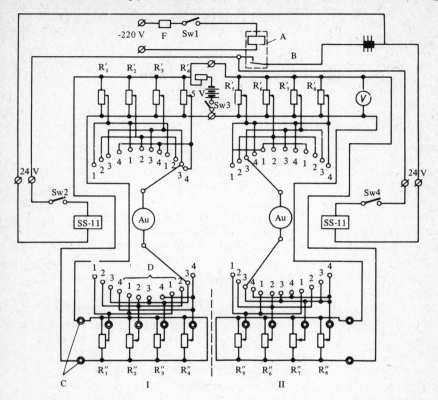

Fig. 13. Measuring layout for the recording of the settlement and displacement deformations of models with the aid of rheostat indicators and self-recording instruments N-370-M.
1-4: fixed contacts of the step selector; I, indicators from the first carriage (package); II, indicators from the second carriage (package); Au, automatic self-recording device N-370-M; SS-11, step selector; R', resistances for zero setting; R", rheostat indicators of deformations (S.I. and D.I.); Sw, switch; F, fuse; A, motor relay of time; B, zero setting; C, current collector rings; D, fixed contacts of the step selector.

The accuracy of the measurements made with the aid of these indicators is as follows: for the S.I. indicators - to 1 mm of settlement; for the D.I. indicators - to one degree of turn of the indicator around its axis of rotation.

The intermediate positions of characteristic points at the brow and foot of a model slope at any moment during the tests are found approximately by diagrammatic plotting on superimposed cross-sectional profiles drawn according to measurements recorded at a given moment and derived from the readings from the S.I. and D.I. indicators at predetermined points of the area and slope of the model, in accordance with the schematic diagram shown in fig. 14.

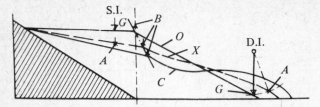

Fig. 14. Schematic diagram for the determination of the intermediate positions of the deforming profiles of the models.

O, initial profile of the model measured by the profilometer; A, the position of the indicators on the intermediate profile being sought; B, positions of the brow of the slope; C, final profile of the model as measured by the profilometer; G, positions of the indicators before commencement of the tests; X, the intermediate profile of the model being sought.

The physico-mechanical properties (and state) of the soil materials of the models (the granulometric composition; the moisture content by weight, w, in per cent; the density expressed by means of the bulk weight of the skeleton, δ, in g/cm^3; and the resistance to displacement φ, in degrees, c in kg/cm^2) are determined by the methods generally adopted in soil mechanics practice and require no additional explanations.

No less than ten separate samples are taken for each model before testing in order to check moisture content, and after the tests five samples are taken at strictly predetermined points in the section of each model as shown in the diagram (fig. 15).

The initial density of the soil mass as deposited in the model is calculated by reference to the overall weight q and volume V of the model and the average moisture content w:

$$\delta_{init} = \frac{q}{V(1+w)} \text{ g/cm}^3.$$

The final densities δ are found for each model by reference to the same test samples used for the determination of the final moisture content of the soil mass after completion of the tests.

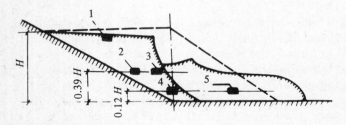

Fig. 15. Diagram showing location of test samples in sloping models. 1–5 are test sample numbers.

Model testing procedure

The granulometric composition, specific gravity, plasticity limits, and also the hardness ratings of the soil mass are determined by duplicated tests throughout the whole series of tests.

In the case of non-sloping models the test samples are taken with a driven core sampler, by means of which one can obtain a column of undisturbed core measuring 73 mm in diameter through the entire height of the model. In such cases the test samples for finding the shear resistance, moisture content and density are taken after each loading (or run on the machine) from each model at strictly predetermined points both in plan and in height.

After the removal of each test sample the cylindrical hole formed in the model is filled by a practically incompressible steel cylinder of identical diameter in order to preserve the integrity of the model during subsequent loadings (runs).

Documentation

The recording of the processes of experimentation with models of waste-heaps proceeds through three stages: primary, intermediate and final documentation.

Primary documentation. This is conducted during preparation and testing of the models. It consists of two logs (preparation and testing of the models), the original recordings of the readings by the automatic recording instruments and the individual matters introduced for each model tested.

In the log of the preparation of the models are noted down the weight of the soil deposited in each model at every loading (or run), the box numbers with the test samples taken for the control determination of moisture content (two test samples for each bucket of soil) and the overall weight of each container complete with models and the indicators installed on them. This log is kept by the laboratory assistant.

In the log for the tests (the log of the operation of the centrifugal modelling machine) are written the date and type of tests, the numbers of the models in order of test, the numbers of the runs (loadings) and the weights of each container (I and II) corresponding to each run, the proposed schedule for the tests, the maximum number of revolutions per minute and duration of the run on the machine, the actual schedule of the tests, the time at the start and conclusion of the tests, the duration of the acceleration runs, periods of constant speed and periods of braking, and also any additional information as required by the experimenter. This log is maintained by the mechanic operating the machine. The original recordings of the automatic recording instruments are preserved, as archive material, on tape-spools (separately for each model).

In the documentation covering the testing of a model there should be the following materials:

1. Records of the determination of the moisture content of all samples of rock or soil taken before and after the tests.

2. Superimposed cross-sectional profiles of the model, taken before the commencement of the tests and after each loading (run) with an indication of the position of characteristic points (the brow, the foot of the slope, a crack); of the surfaces of slip; and of the places where the samples for testing moisture content and density were taken after testing.

3. A plan of the boundary of the surface of slip, taken at a horizontal section at a height equal to 0.417 of the initial height of the model, calculating from the base of the model (fig. 16).

4. A graph showing the running of the model (the loading schedule) in the coordinates t (time) and N (number of revolutions per minute).

5. A decoded interpretation of the auto-recording tape (fig. 17) with the readings of the indicators S.I. and D.I. in the coordinates of settlement Δh, the measurement of the angle $\Delta \theta$ of the rotation of indicator S.I. and the number of revolutions of the model per minute N.

6. An index card of the tests on the model in accordance with the specimen shown in table 5.

The intermediate documentation consists of the materials from the processing of the results of the tests on a group of models. These materials include:

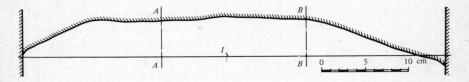

Fig. 16. Plan of a surface of slip (actual measurement on a section of model 13 at a height of $0.417 H$). I is projection of the initial position of the brow of the model slope.

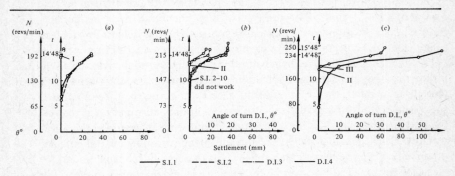

Fig. 17. Interpreted recordings of the readings from the S.I. and D.I. indicators (model 45, runs IV, V, VI).
t = time; a = run IV, b = run V, c = run VI (S.I.2 and S.I.3 did not operate); I = first signs of protrusion; II = resumption of displacement; III = collapse.

Model testing procedure

1. Records of the overall indices, for the whole group of models, of the physico-mechanical properties of the soil materials (granulometric composition, plasticity limits, specific gravity, and hardness ratings).
2. Records of the basic parameters of the slopes of the model waste-heaps at the time of their collapse, compiled separately for each series of models used for the solution of a specific particular problem (determination of the effect on the stability of a waste-heap of moisture content, density of the soil, the rate of deposition and so on); the form of the records is in accordance with the specimen in appendix 1.
3. Calculations.
4. Correspondence on problems concerning the testing of the given series of models.

The final documentation is in the form of an account of the setting up, conducting and results of the tests with the addition of the following documents:
1. Collated records of the characteristics of the soil materials of the models.
2. Test schedule cards.
3. Diagrams showing the ratios $\alpha = f(H)$ for each series of tests and a composite diagram for all the series of tests of the given type.
4. Diagrams showing the sought-for relationship between the critical height of the waste-heap H_{crit} and the variable factors (composition, moisture content, density of the soils, the rate and manner of their deposition in the waste-heap).
5. The resultant data from the calculations carried out.
6. Illustrative material: explanatory sketches and photographs.

The soil materials of the models

In our initial investigations the following soils were taken as standard reference samples: Kerchi tertiary heavy clays, as one of the most hydrophilic soils in the territory of the Ukrainian S.S.R, and Dnieper sands, as an inert, well differentiated, uniform material closest to an ideal starting point for research purposes.

For the modelling of the waste heaps of an (actual) Novy Razdol open-cast mine, soils were taken from the core of two bore holes situated in the northern sector of this open-cast field.

The general physico-mechanical characteristics of the soils used for the models are set out in table 6, and their strength ratings in figs. 18-21.

The soils (apart from sands) used in the preparation of mutually comparable models were subjected to preliminary processing in the form of pulverization in a crusher designed by G. Kh. Pronko (fig. 22) which yields aggregates no greater than 5 mm in size. This corresponds, at an average scale of modelling of $n = 100$, to pieces in the natural state of maximum 0.5 m in size.

The moisture content of the soils when deposited in the spoil-heap in the

Table 5

_____ 19 ___

Index card of tests on

Material of the model _____

Maximum radius of rotation R_{max}

Initial moisture content $w_{av} =$ ___

Indices		Profile number	Height dimensions								
			H_1	B_1	α_1	H_2	B_2	α_2	H_3	B_3	α_3
Loadings (runs)	Measurements										
I	Before test										
	After test										
II	Before test										
	After test										
III	Before test										
	After test										
IV	Before test										
	After test										

Conditions of the commencement and

Commencement of protrusion of the slope											
Commencement of displacement											
Commencement of collapse											

Samplings of the body of

	Final bulk weight of the skeleton, δ (g/cm^3)	Final moisture content w (%)	Notes and additional records

Model testing procedure

model waste heap no. _____

= ____ cm Width of the model l = ____ cm
% Thickness of the sandy base z = ____ cm

U (cm)	Displacement of the brow, (cm)		$R_{eff} = R_{max} - U$, (cm)	Area of section of model F (cm^2)	Volume of the model V (litres)	Weight of the model (kg)	Average bulk weight r_{av} (g/cm^3)	Test schedule				
								Max. number of revs, N (revs/min)	Linear scale of the model	Time of loading, t (min, sec)		
	x	y								Wind-up	Level run

development of the collapse of the model

the model after testing

Table 6 *General physico-mechanical characteristics of the soil materials of the models*

Indices	Unit of measurement	Soils			
		Standard reference		Novy Razdol	
		Kerchi tertiary clays	Dnieper (Kiev) sands	Quaternary sandy loams + clayey loams	Tertiary (Tortonian) clays
Granulometric composition:					
fractions 1.0–0.5 mm	per cent	–	2.2	–	–
0.5–0.25 mm		–	9.0	–	–
0.25–0.05 mm		2.8	50.8	15.3	24.5
0.05–0.01 mm		19.4	34.6	63.0	23.0
0.01–0.005 mm		20.6	1.8	8.1	14.7
<0.005 mm		57.2	1.6	13.6	37.8
Plasticity limits:					
flow limit	per cent	59.4	–	30.9	56.7
plasticity limit		28.6	–	16.6	25.6
Plasticity index	per cent	30.8	–	14.2	31.1
Specific gravity	g/cm^3	2.73	2.66	2.68	27.0
Maximum molecular moisture capacity	per cent	–	3.04	–	–
Angle of repose	degrees	–	31	–	–

natural state is close to the average or average-maximum moisture content of the soils in their naturally deposited state and depends on the lithologic composition, the depth of deposition, the conditions of delivery, and the state of the soil at the time it is tipped on the waste-heap. Hence the moisture content can vary within wide limits; however, it is most frequently to be found close to the limit of total capillaric wetting of the soil when compressed by the appropriate actual load at the working face.

In order to ensure uniform conditions of wetting of the soils placed in the models to be compared it is expedient to take as the basic moisture content that moisture content by weight which corresponds to the total capillaric wetting of the waste soil under some constant compression load. We adopted for this purpose a load with a magnitude of 3.6 kg/cm^2, which corresponds approximately to the compressive load sustained by the soil in the face of a cutting at a depth of 18 m, and also to the load imposed at the base by the self-weight of a waste-heap with a height of 20 metres, given an average bulk weight of the waste soil of 1.8 tonnes/m^3. Pulverised, air-dry Kerchi clay, when loaded in this way, is capable of capillaric saturation with water up to 27–28% of its overall weight (after wetting). For the Dnieper (Kiev) sands the corresponding magnitude has been determined at 13.6%. When modelling actual waste-heaps the moisture content is maintained within the variations observed in the natural state.

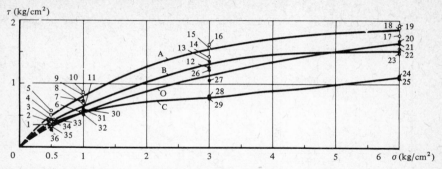

Fig. 18. Shear resistance ratings of Kerchi tertiary clays in a disturbed structure.
O = with capillaric wetting; A = at an initial moisture content of 25.5%; B = the same, at 28.1%; C = the same, at 31.0% moisture content.

1: $(O)w = 32.4; \delta = 1.35$ 2: $(B)w = 28.1; \delta = 1.13$ 3: $(B)w = 28.2; \delta = 1.33$
4: $(A)w = 25.8; \delta = 1.16$ 5: $(A)w = 25.6; \delta = 1.12$ 6: $(C)w = 31.2; \delta = 1.34$
7: $(B)w = 28.5; \delta = 1.42$ 8: $(A)w = 25.2; \delta = 1.21$ 9: $(A)w = 25.6; \delta = 1.28$
10: $(A)w = 25.3; \delta = 1.23$ 11: $(B)w = 27.9; \delta = 1.29$ 12: $(B)w = 28.1; \delta = 1.61$
13: $(B)w = 28.2; \delta = 1.52$ 14: $(A)w = 25.1; \delta = 1.55$ 15: $(A)w = 25.7; \delta = 1.27$
16: $(A)w = 25.6; \delta = 1.35$ 17: $(A)w = 25.6; \delta = 1.56$ 18: $(A)w = 24.9; \delta = 1.68$
19: $(A)w = 25.1; \delta = 1.57$ 20: $(O)w = 24.7; \delta = 1.69$ 21: $(O)w = 24.8; \delta = 1.68$
22: $(B)w = 27.8; \delta = 1.66$ 23: $(B)w = 28; \delta = 1.73$ 24: $(C)w = 31; \delta = 1.55$
25: $(C)w = 30.7; \delta = 1.61$ 26: $(O)w = 26.3; \delta = 1.53$ 27: $(O)w = 27.3; \delta = 1.6$
28: $(C)w = 31.1; \delta = 1.55$ 29: $(C)w = 31.1; \delta = 1.50$ 30: $(O)w = 30; \delta = 1.40$
31: $(O)w = 30.3; \delta = 1.47$ 32: $(C)w = 31.5; \delta = 1.41$ 33: $(C)w = 30.9; \delta = 1.35$
34: $(A)w = 25.1; \delta = 1.02$ 35: $(C)w = 31$ 36: $(O)w = 31.9$
w is in per cent; δ is in g/cm^3.

The moisture content of the soil materials is raised to the predetermined condition, after crushing, with the aid of a water spraying appliance (vacuum dust-collector), after which the rock is kept in hermetically sealed hydrostatic tubs, each with a capacity of 100 litres (fig. 23) for not less than 24 hours before being deposited in the models.

The influence on the stability of waste-heaps of the initial density of deposition or the coefficient of the original degree of pulverisation of the soil in the waste-heap has been disclosed by L. P. Markovich [8], but not entirely correctly interpreted.

In our investigations three gradations of the density of deposition of the soil in the models were adopted for models to be compared: loose deposition, by pouring the soil freely from a shovel, when minimum density is created, expressed via the bulk weight of the skeleton of the soil as δ_{min}; dense deposition, with maximum possible consolidation by means of ramming (to constancy of volume), when δ_{max} is achieved; and deposition with average density:

$$\delta_{av} = \tfrac{1}{2}(\delta_{min} + \delta_{max}).$$

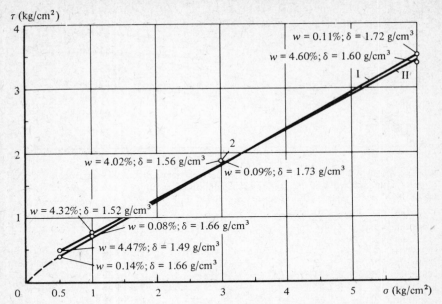

Fig. 19. Shear resistance ratings of Dnieper sand (from the shore of the Kiev reservoir): I = practically dry sand; II = with moisture content $w \approx w_M$.

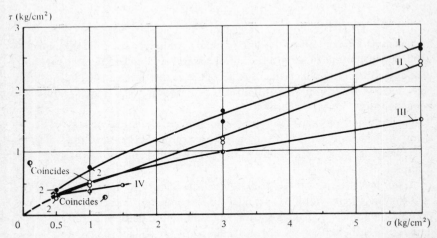

Fig. 20. Shear resistance ratings of the soil materials of the model waste-heaps consisting of quaternary soils (sandy loam and clayey soil) of a Novy Razdol deposit: I, $w = 20\%$; II, with capillaric wetting; III, $w = 22.5\%$; IV, $w = 25\%$.

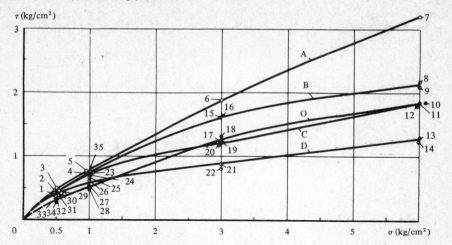

Fig. 21. Shear resistance ratings of Razdol tertiary clays in a disturbed structure.

O, with capillaric wetting; A, at an initial moisture content of 18.8%; B the same, at 28.2%; C, the same, at 30.9%; D, the same, at 34.0%.

1: $(C)w = 30$; $\delta = 1.1$	2: $(D)w = 34.6$; $\delta = 1.2$	3: $(A)w = 22.1$; $\delta = 1.09$
4: $(C)w = 29.8$; $\delta = 1.24$	5: $(A)w = 19.2$; $\delta = 1.16$	6: $(A)w = 19.2$; $\delta = 1.44$
7: $(A)w = 19.4$; $\delta = 1.53$	8: $(B)w = 26.6$; $\delta = 1.61$	9: $(B)w = 26.5$; $\delta = 1.59$
10: $(O)w = 25.4$; $\delta = 1.65$	11: $(C)w = 27.8$; $\delta = 1.72$	12: $(C)w = 27.9$; $\delta = 1.66$
13: $(D)w = 29.6$; $\delta = 1.51$	14: $(D)w = 29.8$; $\delta = 1.52$	15: $(B)w = 26.8$; $\delta = 1.46$
16: $(B)w = 26.8$; $\delta = 1.47$	17: $(O)w = 27.7$; $\delta = 1.56$	18: $(O)w = 26.2$; $\delta = 1.58$
19: $(C)w = 29.8$; $\delta = 1.54$	20: $(C)w = 29.8$; $\delta = 1.56$	21: $(D)w = 32.6$; $\delta = 1.39$
22: $(D)w = 32$; $\delta = 1.45$	23: $(B)w = 27.3$; $\delta = 1.19$	24: $(C)w = 29.8$; $\delta = 1.24$
25: $(D)w = 33.8$; $\delta = 1.29$	26: $(D)w = 33.2$; $\delta = 1.26$	27: $(O)w = 31$; $\delta = 1.51$
28: $(O)w = 31.2$; $\delta = 1.31$	29: $(B)w = 27.4$; $\delta = 1.12$	30: $(B)w = 27.1$; $\delta = 1.08$
31: $(D)w = 34.2$; $\delta = 1.11$	32: $(C)w = 29.6$; $\delta = 1.17$	33: $(O)w = 31.2$; $\delta = 1.39$
34: $(O)w = 31.4$; $\delta = 1.45$	35: $(B)w = 27.3$; $\delta = 1.19$	

w is given in per cent; δ is in g/cm³.

The magnitude H_{crit} and the ratio $\alpha = f(H)$ are found in cases of other intermediate values of δ by interpolation.

For the Novy Razdol open-cast mines L. P. Markovich [8] determined the limit values of the primary coefficient of bulking (increase in bulk volume) of the soil at the first moment of its deposition in the waste-heaps as follows:

For quaternary sandy loams and clayey loams, 1.41-1.23

For tertiary clays, 1.43-1.21

If we take into account the fact that the bulk weights of those same soils in their undisturbed state were determined by him as: for the first case 2.04 t/m³ and for the second 2.05 t/m³, then the bulk weight of the soils deposited in the waste-heaps will be determined as 1.44-1.66 t/m³ for quaternary sandy loams and clayey loams and as 1.45-1.69 t/m³ for tertiary clays. The bulk weight of the soil as shown in table 7 corresponds to these values, at various degrees of moisture content of the soil.

Fig. 22. Soil-crushing and grinding machine designed by G. Kh. Pronko. 1, crushing plate; 2, grinding device.

Model testing procedure

Fig. 23. Hydrostatic tubs for storing prepared rock materials.

It is difficult to observe and maintain these conditions when building a model. Therefore for the models of the Novy Razdol waste-heaps also we retained the conditions of modelling using δ_{min}, δ_{max} and $\delta_{av} = \frac{1}{2}(\delta_{min} + \delta_{max})$.

The allocation of operatives

The rational organisation of the centrifugal modelling of waste-heaps requires the coordinated work of a team of six operatives with the following allocation of responsibilities.

1. Leader of the investigations – organisation and coordination of all work, development of programmes and investigation procedures, calculations and processing of the results of the investigations.

2. Senior engineer – the deputy of the leader – preparation and measurement of the models, calculations and processing of the results of the investigations, deputy for the senior laboratory assistant.

3. Mechanic supervising the centrifugal assembly – centrifuging experiments according to the set programme, operational service of the material aspects of the centrifugal modelling installation, the design and construction of secondary instruments and apparatus, participation in the preparation of the models.

4. Tensometry engineer – organisation and realisation of all remote observations and measurements and their processing, participation in the preparation and testing of the models, deputy for the chief mechanic.

5. Senior soil mechanics laboratory assistant – preparation of the soil materials

Table 7 Initial density of the soil mass, expressed in terms of the bulk weight of the skeleton, in Novy Razdol waste heaps and their models

Soils	As determined by L. P. Markovich			Bulk weight of the soil in the waste-heap (initial) $\gamma_0 = \gamma_u/K$ (t/m³)		Initial bulk weight of the skeleton of the soil in the waste-heap				Bulk weight of the skeleton of the soil materials used in the model (t/m³)		
	Average bulk weight of the untouched soil γ (t/m³)	Primary coefficient of granulation when soil deposited in the waste-heap, K				At moisture content w (per cent)	$\delta = \dfrac{\gamma_0}{1+w}$ (t/m³)					
		From To	Average value	From To	Average value		From To	Average value	δ_{min}	δ_{max}	δ_{av}
Quaternary sandy loams and clayey loams	2.04	1.41 1.23	1.32	1.44 1.66	1.55	20 22.5 25	1.20 1.38 1.17 1.36 1.15 1.33	1.29 1.26 1.24	0.91	1.54	1.22
Tertiary clays	2.05	1.43 1.21	1.32	1.45 1.69	1.56	25 28 31	1.16 1.35 1.13 1.32 1.11 1.29	1.25 1.22 1.19	0.75	1.21	1.00

for construction of the models, laboratory determinations of their physico-mechanical and strength properties, participation in the preparation, testing, measurement and dismantling of the models.

6. Assistant – preparation of soil materials for the models, their crushing, wetting, mixing, auxiliary work in the preparation and dismantling of the models.

2

The accuracy of the modelling of unconsolidated waste-heaps

The limits of the correspondence of models to the conditions of similarity for unconsolidated waste-heaps were subjected by ourselves to a double check by comparing the results of modelling using models of different scales and by comparing the results of modelling known waste-heaps in the natural state with the results occurring in the actual natural conditions.

In the first case models of different scales, prepared in identical manner, from identical material and with identical initial moisture contents and densities, and tested under equivalent loading procedures, were brought to a state of collapse by slipping whilst recording on a time scale the deformations of settlement and displacement of the slope. Comparisons of the results of the tests, when converted to the scale of the 'natural state', were made according to the features of the geometrical similarity of the deforming cross-sectional sections of the models and by means of juxtaposition and comparison of the degree of consolidation of the mass of the waste-heap, and the change in its initial moisture content and strength at various equivalent points in the cross-sectional sections of the models.

In the second case, in the management of natural waste-heaps, it is impossible to maintain the constancy of the initial moisture contents and densities of the soils deposited in the waste-heaps. In place of the constancy of these indices it was necessary to make use of the measurements of the observed limits of their fluctuations and to carry out the modelling in series of tests within the limits of the corresponding ranges with the subsequent construction of curves showing the relationship between height and moisture content and density.

The investigation of sloping models

An essential condition of the centrifugal modelling of waste-heaps is the maintenance of similarity of the distribution of loads in comparable models at any moment during the increase in accelerations. The type of cross-sectional section of a sloping model waste-heap which we adopted and which has been described above (see fig. 6) satisfies this condition.

In the present case comparisons are made of the results of the tests on three

Accuracy of modelling

series of model waste-heaps with initial heights of 24, 30 and 36 cm. There are four models in each series. Each model has two fixed sections (A and B) in the middle third of the width for the measurement of the deforming profiles both during and at the end of the tests. The numbers and initial dimensions of the models are shown in table 8. The internal dimensions of the widths of both model containers do not correspond equally to each other. Hence the initial volumes of the pairs of models being tested simultaneously differ slightly from each other.

At the start-up of the centrifugal machine the containers turn through almost 90°. The distances R_0 from the axis of rotation to the bottom of both containers vary somewhat. For container I, $R_0^I = 2.588$ m, for container II, $R_0^{II} = 2.591$ m. The effective radii ($R_{eff} = R_0 - y_0 - z$) of the initial sections of the models undergoing testing are set out in table 9.

All twelve models were prepared from Kerchi tertiary clays. Their physico-mechanical characteristics are given in table 6 and in fig. 18.

The technology and norms for the preparation of the soil for all the models were identical. Where substantial deviations from them occurred the model was rejected entirely or in part.

Despite the care in the preparation of the tests we did not succeed in maintaining a constant initial moisture content and density for all twelve models,

Table 8 *Initial dimensions of the comparative sloping models*

Numbers of the models	Linear dimensions (cm) (see fig. 6)					Sectional area F (cm²)	Initial volumes of the models, V (litres)	
	H	B	D	y	z		Even numbers	Odd numbers
8, 9, 12, 13	24.0	33.0	41.6	12.5	5.5	895	43.6	44.0
10, 11, 14, 15	30.0	41.3	52.0	15.6	5.5	1400	68.4	69.0
6, 7, 16, 17	36.0	49.5	62.4	18.7	5.5	2010	98.1	99.0

Table 9 *The effective radii of the models before commencement of their deformations*

Numbers of the models	Effective radii (metres)	
	Without rounding off	Rounded off
8 and 12	2.408	
9 and 13	2.411	2.41
10 and 14	2.377	
11 and 15	2.380	2.38
6 and 16	2.346	
7 and 17	2.349	2.35

which had an overall total volume in excess of 2500 litres of loose mass. In table 10 are shown the deviations obtained from the accepted average initial values, which were: moisture content by weight w_{av} = 27.29%, and average density of loosely deposited soil, expressed in terms of the bulk weight of the skeleton, δ_{av} = 0.664 t/m³.

The tests were conducted at final scales of modelling ranging from n = 95 to n = 265 with intermediate measurements of the profiles of the models on each occasion the centrifugal machine was stopped, starting from scales of n = 25, n = 30 and n = 37.5 at the pre-established schedules of loading (runs) set out in table 3.

After each loading (run) the machine was stopped and the cross-sectional profiles were measured across predetermined sections of the model, after which fresh soil material was poured on to restore the model to its original height. After that the model was subjected to fresh loading (and a fresh run) up to the subsequent scaled degree.

The deformations of models of clayey waste-heaps always occur in an identical sequence, which is well traced on superimposed profiles (see fig. 11). In the beginning the loose soil mass in the model is intensively consolidated, and this is accompanied by settlements combined with flattening of the angle of the slope only in the upper part and the horizontal displacements of the upper layers in the direction of the slope. When this occurs, stepped protrusions appear on the slope, which appear to be rolling down in waves one over the other. The lowest wave of the protrusion remains within the original contours of a slope, formed at the natural angle of repose of the tipped material.

Table 10 *Initial moisture content and density of the soil materials used in the models*

Initial height of the model H_M (cm)	Model number	Initial indices	
		Moisture content, w (%)	Density δ (t/m³)
24	8	27.87	0.686
	9	27.90	0.688
	12	27.25	0.686
	13	27.39	0.676
30	10	27.34	0.675
	11	27.60	0.672
	14	27.12	0.669
	15	27.12	0.662
36	6	26.63	0.608
	7	26.70	0.641
	16	27.20	0.668
	17	27.36	0.643
		Average values:	
		27.29	0.664

Accuracy of modelling

During subsequent loadings (runs) the height of the lowest wave of the protrusion is gradually raised till it reaches a certain limit approaching the height of the brow of the waste-heap which is being deposited and not extending beyond the contours of a slope inclined at the angle of repose. When the limit which corresponds to the moment of achievement of the waste-heap's critical height is exceeded, a bulge in the lower part of the waste-heap begins to protrude beyond the original contour of the slope. Simultaneously, surfaces of slip are formed in the body of the clayey waste-heap which, when fully developed, take the form of a continuous curvilinear slickensided rupture, and displacement of the slope occurs, accompanied by the fissuring and crumpling of the soil in the slipping body.

In our case the surface of slip broke the surface of the upper area of the waste-heap at a distance, on average, of $0.25\,H$–$0.26\,H$, calculating from the brow of the slope, and was approximately roundly cylindrical in shape.* In the majority of models of clayey waste-heaps a single unbroken surface of slip is formed. But in rare cases we observed two parallel (in the curved part) surfaces of slip which merged into one in the lower levels. As it descends to the foot of the slope the surface of slip is converted into a plane, parallel to the firm sandy foundation and at a distance of 1-2 mm from it.

Test samples of soils taken from the surfaces of slip reveal in the majority of cases a tendency towards a certain increase in moisture content at the surfaces of slip by comparison with the moisture content of the adjacent bulk masses of the waste-heap material. Comparative determinations of the moisture content of the soils at the surfaces of slip and in the soil masses adjacent to them are shown in table 11.

Table 11 *Comparison of the moisture content of the soil materials of the models at the surfaces of slip and in the soil masses adjacent to them*

Model number	Moisture content by weight, w (%)						Difference between the values of w (%)
	At the surface of slip			In the adjacent parts of heap			
	From	To	Average value	From	To	Average value	
6			27.95			28.10	−0.15
8	29.02	29.40	29.21	27.30	27.32	27.31	+1.90
9	26.85	27.34	27.10			28.32	−1.72
10	25.90	26.42	26.16			25.40	+0.76
12	27.50	27.30	27.40	27.35	27.10	27.22	+0.18

*The extreme limits were $0.14\,H$ and $0.42\,H$.

During subsequent tests with models containing predominantly sandy loam material it became clear that the character of the course of their deformations differs significantly from the deformations of clayey waste-heaps. The formation of wave-like protrusions on their slopes is sometimes less sharply expressed. The slopes collapse in multi-stage escarpments, without the formation of a single surface of slip. This gives the collapse the appearance of having been torn away, with many abrupt fissures and shallow slip terraces at the surface.

The absolute linear dimensions of models with different initial heights are not comparable. Hence all comparisons of the results of tests can be effected only after conversion to the scale of an equivalent 'natural object'. The comparisons are made on the basis of both fundamental and secondary indices.

We take as fundamental the indices of the stability of an equivalent waste-heap slope, expressed in terms of $H_{\text{crit N}}$ and in terms of the relationship $\alpha_{\lim} = f(H_N)$.

We assign to the category of secondary indices the other magnitudes of the geometrical similarity of the models before the commencement of their collapse: the height of the lower terrace of protrusion h_N, the average angle of slope within the limits of the terrace of protrusion $\alpha_{\max}$, and also the indices of the physical state of the soil mass—moisture content w, density δ and strength φ, c in equivalent points of the models.

The final comparative determinations, after conversion to the scale of an equivalent 'natural object', of the critical height of a waste-heap (H_{crit}) and the relationship $\alpha = f(H)$ are presented in the graphs in fig. 24.* The results of the investigations of the limit of the mutual correspondence of models of different scales were subjected to analysis by the method of mathematical statistics (appendix 2). The mean square error when determining H_{crit} = 38.96 m did not exceed 1.22 m at the maximum coefficient of variation 0.033. The maximum error (where P = 0.95) was equal to 2.62 m or 6.8%.

The deviations in the determinations of the limit angles of slope, $\alpha_{\lim}$, in the range of heights of the waste-heaps from H_{crit} up to 50 m, did not exceed 1°12'.

A check on the magnitudes of the differences in the averages for models of various heights, carried out by the method of dispersion analysis (appendix 3) revealed an absence of any significant difference between them. Any observable deviations are found to be within the boundaries created by random variation of the data. In other words, on the basis of the experiments conducted in the volume reduced one cannot draw any well-founded conclusion concerning the existence of a substantial difference between the results of tests in models of different initial heights. The observed variations in the average values between different groups of models are accounted for by accidental reasons, i.e. by the influence of factors not taken into account. The determinations of the reduced height of the waste-heaps at the recorded brows of the slopes in all stages of

---

*An example of the registration of the basic parameters of the slopes of model waste-heaps during their collapse is shown in the table of appendix 1.

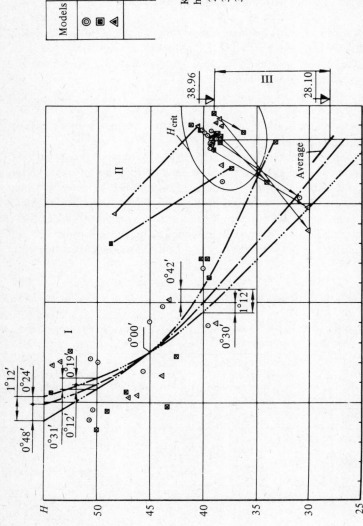

Fig. 24. The relationship $\alpha_{lim} = f(H)$ and the critical values of the height of a waste-heap H_{crit} for models with different values of initial height (Kerchi tertiary clays, $w_{av}^{init} = 27.29\%$, $\delta_{av}^{init} = 0.66$ g/cm³).

I, the relationship $\alpha_{lim} = f(H)$ after formation of the surface of slip; II, before the formation of the surface of slip; III, the average range of the reduction in the critical height H_{crit} after the formation of surfaces of slip in the models ($w_{init} \approx 28.0\%$).

loading up to the onset of the collapse of the models vary within approximately the very same limits (table 12).

More significant deviations are encountered (see table 12) when determining the height of the lower terrace of protrusion h_N. This may be explained by the conditional nature of the magnitudes, since the places on the models where they are measured are by no means fixed and are designated by the judgement of the observer within the boundaries of the transition (sometimes fairly even) from one wave on the surface of the slope to another more slanting wave. In this circumstance, the variations in the measurements naturally turn out to be greater (whereas the position of the brow of the model is fixed by datum marks which are displaced together with the brow of the slope during deformations of the model).

Despite the insignificance of the deviations in the determinations of the basic indices of stability we can discover that with a reduction in the scale of the models there is a tendency towards an increase in the results obtained for the height of the waste-heaps H_N (fig. 25).

$$H_N = H_M n (1 + k_{av}),$$

where n is the scale of modelling; $k_{av} \approx -0.003$ - a correction, which we shall in future disregard.

The results of sample checks, after the tests, of the material of the models for moisture content w and density δ are set out in table 13. The considerable variations of both indices in the upper levels of the models (test sample no. 1) are explained by the imperfection of the method of taking samples of loose soil with the aid of a measuring cylinder, which did not permit us to achieve constant filling of the cylinder, and by the lack of uniformity in the losses of moisture content at the surface of the models during the lengthy process of dismantling and cutting up

Table 12   *Comparison of the magnitudes H_N, α°_{res}, h_N and α°_{max}, found in models of varyi*

Loadings (runs)	Scales of modelling, n, for models with initial height, $H_{M\,init}$ (cm)			Height of waste-heap at the brow, H_N (metres)				Resulting angle of slope α_{res}		
				Average of eight measurements for models of initial height $H_{M\,init}$ (cm)			General average (by reduction)	Average of eight measurements for models with initial height $H_{M\,init}$ (cm)		
	24	30	36	24	30	36		24	30	36
I	37.5	30	25	6.26	6.35	6.31[a]	6.32	28°42'	28°57'	28°15'
II	75	60	50	13.8	14.0	13.8[a]	13.9	30°16'	31°03'	30°00'
III	112.5	90	75	21.9[b]	22.1	21.7[a]	21.9	31°11'	32°05'	31°06'
IV	150	120	100	30.4[c]	30.6	29.8[a]	30.1	32°19'	33°04'	32°01'

[a] Average of four measurements.   [b] Average of six measurements.   [c] Average of seven measurements.

Accuracy of modelling

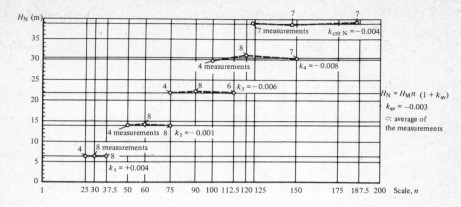

Fig. 25. The scale factor in the determinations of H_N and H_{crit} in models.

the models. All the remaining test samples (nos. 2-5) were taken from the depth of the models. Their bulk weights were determined by the paraffinisation method. In these cases the coefficients of variation in the determination of the values of δ are found within the limits 0.005-0.024, and of the values of w within the limits 0.012-0.042.

The test samplings of the strength of the body of the models by means of a hand-operated penetrometer, introduced horizontally into the body of the model at the places of intersection of grid lines, drawn on the side surfaces of the models, did not yield satisfactory results.* The coefficients of the variations in the determinations with this kind of test sampling will fluctuate from 3.9% to 16%. Such a low degree of accuracy is explained by the imperfection of the

*Dimensions of grid $0.139 H \times 0.139 H$.

itial height under four loadings (before the onset of collapse of the models)

	Height of the terrace of protrusion h_N (metres)				Average angle of slope within the confines of the terrace of protrusion α_{max}			
:neral erage y reduction)	Average of eight measurements with initial height $H_{M\ init}$ (cm)			General average (by reduction)	Average of eight measurements for models with an initial height $H_{M\ init}$ (cm)			General average (by reduction)
	24	30	36		24	30	36	
°38′	2.8	2.2	2.0	2.3	33°20′	31°44′	33°04′	32°43′
°26′	6.2	6.4	5.9	6.2	33°02′	34°30′	32°58′	33°30′
°27′	12.9	12.9	11.6	12.4	33°54′	35°09′	33°54′	34°20′
°28′	23.3	21.8	21.7	22.3	34°42′	35°09′	34°53′	34°55′

Table 13 *Results of the test-sampling of soil materials from sloping models after*

Test number	Average initial moisture content (%)					Actual final moisture-content				
	Model numbers, having an initial height H_{init} = 36 cm									
	6	7	16	17	Weighted average	6	7	16	17	A· va
1						27.7	26.4	25.3	26.1	26
2	26.6	26.7	27.2	27.4	26.9	26.7	25.7	25.7	26.4	26
3	From 15 tests	From 15 tests	From 15 tests	From 16 tests		25.8	25.2	25.7	26.3	25
4						26.8	26.7	25.3	26.0	26
5						25.3	26.8	26.1	27.2	26
	Model numbers, having an initial height H_{init} = 30 cm									
	10	11	14	15		10	11	14	15	
1						26.0	24.7	25.0	25.0	25
2						25.7	25.2	26.8	26.3	26
3	27.3	27.6	27.1	27.1	27.3	26.0	26.8	26.7	27.3	26
4	From 20 tests	From 21 tests	From 18 tests	From 21 tests		26.3	26.3	27.4	27.2	26
5						26.2	26.1	28.0	28.0	27
	Model numbers, having an initial height H_{init} = 28 cm									
	8	9	12	13		8	9	12	13	
1						25.4	25.7	26.2	25.8	25
2						25.8	25.6	26.1	25.8	25
3	27.9	27.9	27.3	27.4	27.6	25.8	26.2	26.7	26.4	26
4	From 18 tests	From 18 tests	From 18 tests	From 17 tests		26.0	26.0	27.4	26.1	26
5						26.4	27.4	28.0	27.0	27

method of determination itself, in which the readings on the scale of the penetrometer permit errors up to 6.3%, and the accidental features of the penetration of the penetrometer into dense lumps of soil in one place and into fissures between such lumps in another make it impossible to achieve any uniformity of measurement. This obliged us to reject the method of penetration test sampling as applicable to the non-homogeneous bodies of model waste-heaps and to transfer to more sophisticated methods of determining the strength of the waste-heap mass using shear stress instruments.

The investigation of non-sloping models

On account of the presence of a slope in a model waste-heap local concentrations of stresses arise, and also plastic shear and rupturing deformations which lead to uncontrollable changes of density, strength, and sometimes also of the moisture content of the soil in the body of a model waste-heap. Therefore we additionally

Accuracy of modelling

ysis for moisture-content w and density δ

Reduced (to $w_{init} = 27\%$) final moisture-content (%)					Bulk weight of the skeleton δ (g/cm³)				
6	7	16	17	Average value	6	7	16	17	Average value
28.1	26.7	25.1	25.7	26.4	1.10	1.10	1.08	1.10	1.09
27.1	26.0	25.5	26.0	26.1	1.51	1.54	1.53	1.54	1.53
26.3	25.5	25.5	25.9	25.8	1.53	1.53	1.52	1.52	1.53
27.2	27.0	25.1	25.6	26.2	1.52	1.47	1.55	1.53	1.52
25.7	27.1	25.9	26.8	26.4	1.54	1.52	1.54	1.53	1.53
10	11	14	15		10	11	14	15	
25.7	24.1	24.9	24.9	24.9	1.08	1.12	1.07	1.03	1.08
25.4	24.6	26.7	26.2	25.7	1.53	1.55	1.51	1.54	1.53
25.7	26.2	26.6	27.2	26.4	1.55	1.54	1.52	1.51	1.53
26.0	25.7	27.3	27.1	26.5	1.55	1.56	1.53	1.55	1.55
25.9	25.5	27.9	27.9	26.8	1.50	1.52	1.51	1.52	1.51
8	9	12	13		8	9	12	13	
24.5	24.8	25.9	25.4	25.1	1.13	1.19	1.09	1.20	1.15
24.9	24.7	25.8	25.4	25.2	1.54	1.55	1.53	1.53	1.54
24.9	25.3	26.4	26.0	25.7	1.50	1.54	1.51	1.52	1.52
25.1	25.1	27.1	25.7	25.7	1.51	1.55	1.53	1.55	1.54
25.5	26.5	27.7	26.6	26.6	1.52	1.51	1.52	1.52	1.52

checked for the existence of concurrence in the results of test samplings in models of unconsolidated waste-heaps, according to those criteria, in that part of the waste-heap where the influence of the proximity of a slope is excluded, i.e. where only the normal compressive stresses are acting and where the changes in the physico-mechanical properties and the state of the soil of the waste-heap occur only as the consequence of the consolidation of the soil under the influence of its own gravity.

The non-sloping models nos. 18 and 19, with an initial height of 24 cm, were also tested. These were totally identical in their material composition, technique of deposition and methods of loading with the sloping models nos. 8, 9, 12 and 13, which had the same initial height.

The dimensions of the non-sloping models in plan were fixed so as to permit the removal of test-samples in the form of a continuous core through the whole depth of the models after each loading (run) of the model and at the same time so as to eliminate the retarding effect on settlement of the walls of the containers and of the steel cylindrical plugs used to fill up the holes in the models left after removal of the core-samples. The diameter of the core samples which were taken was 74 mm; the external diameter of the sampling tube was 82 mm.

According to the limits of 20 sloping models the established mean distance of the retarding influence of the fixed wall of the container is 11.4 cm.

The schemes for the arrangement of, and the order for the removal of, the core samples which we adopted for the plan section of the model (see fig. 26) satisfy the condition of preventing any retarding influence of the adjacent fixed walls and of the plugs.

The distribution of the test samples through the vertical section of the model, through the continuous core-sample removed by the sampling tube, was arranged so that they intersected the shear-test samples along planes at strictly fixed marks. In table 14 the conditions are set out for the division of the test-samples of non-sloping models along the vertical axes, which we adopted for the following reasons.

In fig. 27 are shown the curves of the average values of the settlement of the horizontal surfaces of sloping models as determined at the crests and at 3.3 cm from the crests (where the settlement sensors were located) after each loading (or run) of the models. The average values were derived from measurements on eight sections of the sloping models.

It was anticipated that the settlements of the non-sloping models would be close to the average between the two curves. Table 15 shows the calculated, anticipated magnitude of the settlements and of the corresponding final heights of the models after each of five loadings (runs) before the onset of the collapse of the sloping models.

The loading ring of the Maslov shear device has a height of 3.5 cm. Leaving 0.5 cm for the slicing and trimming of the samples, it is possible to obtain from every four centimetres of height of core removed by the tube-sampler one test-sample for the determination of the shear-resistance, moisture content and bulk

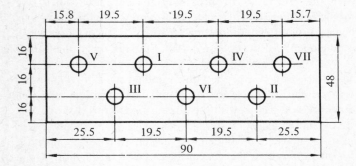

Fig. 26. Diagram showing the disposition in plan and the order of removal of continuous core-samples from non-sloping models. Measurements are in centimetres.

Table 14 *Conditions for the vertical splitting of non-sloping models, for the removal of test-samples*

Distances between the tops of the loading rings	Diagram of the division of the core-sample column	Marks for the cutting planes (cm)	Scale of modelling, n	Equivalent depth of the cut in the waste-heap (metres)	Normal loading on the surface of the cut σ (kg/cm^2)	Scale of modelling, n	Equivalent depth of the cut in the waste-heap (metres)	Normal loading on the surface of the cut σ (kg/cm^2)	Scale of modelling, n	Equivalent depth of the cut in the waste-heap (metres)	Normal loading on the surface of the cut σ (kg/cm^2)
			I Run/loading			*II Run/loading*					
0.5		0									
4.0		−2.5	37.5	0.94	0.10	75	1.86	0.20			
4.0		−6.5		2.44	0.28		4.87	0.62			
4.0		−10.5		3.94	0.48		7.86	1.14			
4.0 +x		−14.5		5.45	0.71		10.88	1.64			
			III Run/loading			*IV Run/loading*			*V Run/loading*		
0.5		0									
4.0		−2.5	112.5	2.84	0.33	150	3.75	0.46	187.5	4.68	0.60
4.0		−6.5		7.31	1.01		9.75	1.44		12.20	1.88
4.0		−10.5		11.82	1.81		15.78	2.55		19.70	3.30
4.0		−14.5		16.31	2.66		21.75	3.70		27.20	4.75
4.0 +x		−18.5		20.80	3.51		27.75	4.86		34.70	6.20

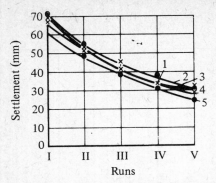

Fig. 27. Anticipated and actual settlements of non-sloping models nos. 18 and 19.
1, average indices from 8 sensor devices; 2, average settlements of the crests of 4 sloping models with H_{init} = 24 cm; 3, average actual settlements of non-sloping models nos. 18 and 19; 4, expected settlements of non-sloping models with H_{init} = 24 cm; 5, average surface settlements of 4 sloping models with H_{init} = 24 cm.

Table 15   *Calculation of the anticipated settlements of non-sloping models*

Indices	Serial numbers of the loadings (runs)				
	I	II	III	IV	V
Anticipated settlement of model (cm)	6.6	5.1	4.1	3.4	2.7
Anticipated residual height of the model H_M (cm)	17.4	18.9	19.9	20.6	21.3
Scale of modelling	37.5	75	112.5	150	187.5
Equivalent height of waste-heap H_N (m)	6.25	14.18	21.40	30.90	40.00

weight of the skeleton of the soil at each given mark on the model. Thus a core with a height of not less than 16.5 cm yields four test-samples, and one with a height of 20.5 cm – five test samples.

The core is divided up by cutting in the first and second cases, starting from the top in accordance with the diagrams in table 14. In those same diagrams are indicated the places where slits are made and also the corresponding calculated normal loadings, equal to the maximum compressive loadings sustained at the given mark (level) by the model during centrifuging.

The average bulk weight of loosely deposited soil in models nos. 8, 9, 12 and 13 was equal to $\gamma = 0.873$ g/cm^3.

The bulk weight of the soil mass after the centrifuging of a sloping model with an initial height of 24 cm and moisture content w = 27% at depths of 1.75, 12.5 and 18.8 cm at a scale of modelling n = 187.5 was provisionally determined as equal to 1.41, 1.93 and 1.94 g/cm^3 respectively. After conversion to the equivalent heights of a natural waste-heap at the four points established, a

Accuracy of modelling

provisional graph (for calculation purposes) of the changes in the bulk weight of the soil though the depth of the waste-heap which had been subjected to loading was then constructed (fig. 28).

The estimated normal loadings σ, adopted in shear-resistance tests, were defined as the overall weight of the layers of soil overlaying the plane of shear:

$$\sigma = \gamma_1 h_1 + \gamma_2 h_2 + \cdots + \gamma_n h_n,$$

where γ_i is the average bulk weight of a layer of soil i; h_i is the thickness of that same layer.

Since after each loading (run) of the model it is topped-up to its original height, the upper layers repeat in each new loading the conditions of the preceding one, but in another scale of modelling.

As a result of five loadings (runs) in two models 23 samples were obtained for each, permitting comparison both between themselves and with duplicated samples of the paired model.

The results of the tests on non-sloping models are set out as follows: distribution of moisture content w in the body of the models and changes in the density of the soil mass according to height in the models – table 16 and fig. 29;

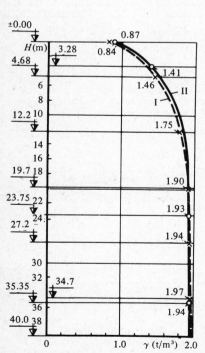

Average-weighted bulk weights γ (tonnes/m³)		For depths H (metres), from–to
Calculated	Measured	
1.10	1.00	0–2
1.36	1.30	2–4
1.52	1.48	4–6
1.66	1.60	6–8
1.76	1.70	8–10
1.82	1.77	10–12
1.86	1.82	12–14
1.90	1.88	14–18
1.92	1.90	18–22
1.93	1.94	22–32
1.94	1.97	32–40

Fig. 28. Graph of the relationship $\gamma = f(H)$. I = average according to measurements on non-sloping models; II = calculated.

shear resistance of the body of the models at various depths underground – table 17 and fig. 30.

The scatter of points for the determination of the moisture content of the

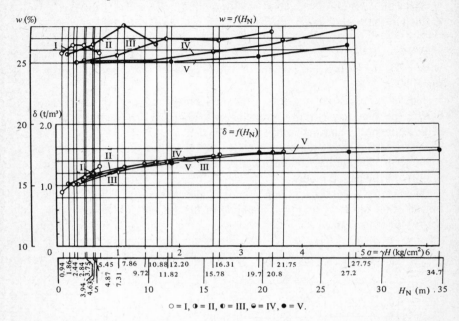

Fig. 29. The distribution of moisture content $w = f(H_N)$ and changes in the density of the soil mass δ in the body of a non-sloping model waste-heap, converted as applicable to the height of the waste-heap being modelled, H_N. I–V are the different runs.

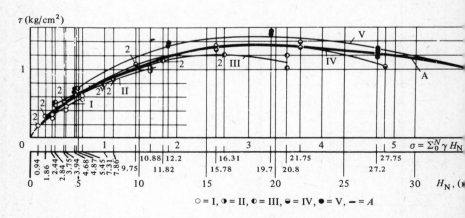

Fig. 30. Graph showing the relationship $\tau = f(H_N)$ for soils from a model of an unconsolidated waste-heap where $\sigma = \Sigma_0^N \gamma H$ kg/cm², A is the average of five runs; I–V are the different runs.

body of the models in the given case, in contrast to the preceding tests with sloping models, varied within a 3% range (25–28%), revealing a tendency towards an increase in moisture content in the lower levels of the model, which was not confirmed by a quantitative check in subsequent tests and was evidently accidental.

The concurrence of the results of the determination of density δ and shear-resistance at various points in the body of the waste-heap, at various scales of modelling, was satisfactory and did not exceed the coefficients of variation determined for the basic indices of the stability of waste-heaps.

The density of the soil mass of a waste-heap comprising Kerchi tertiary clays increases intensively to a depth of 15–18 m, but the intensity falls and from a depth of 20 m onwards (under compressive loads greater than 3.5 kg/cm^2) the density becomes almost constant. At a depth of from 20 to 30 m the bulk weight of the skeletal soil mass increases only from 1.50 to 1.56 t/m^3, i.e. the pores of the mass of waste material are almost completely closed. On the graph showing the relationship $\tau = f(H)$ this moment corresponds to the maximum shear resistance of the soil mass. With a further increase in the compressive loads the shear resistance of the clayey soil mass begins to fall and when the normal loads corresponding to depths of 55–60 m (of the order of 10.5 kg/cm^2) are reached, at a moisture content of the soil of 25–27%, a fall in the shear resistance of these soils to very low levels should be expected.

At great depths a similar transition of the clayey mass into the state of a viscous liquid is possible even with a lower water content. Such phenomena occur when underground workings encounter self-discharging streams of clay or those which spill out as fountains.

We find the most probable explanation of the physical significance of the fall in the resistance capacity of clays with an increase in compressive loadings in the development of excess pore pressure from the water, starting from the point corresponding to the conditions of transition of the soil mass from a three-phase to a two-phase state.

In the case cited all the shear tests were carried out on samples removed immediately after centrifuging and sheared under normal loadings, equal to those maximum compressive loads to which the samples had been subjected during the centrifuging process.

An interesting phenomenon was observed when shear tests were carried out on unconsolidated soils from models nos. 18 and 19, taken from various depths after the fifth (last) loading (run) and sheared under a constant compressive loading $\sigma = 3.5$ kg/cm^2, which is close to that at which, it is presumed, excess pore pressure from the water arises. The results of these tests are set out in table 18 and in fig. 31.

The moisture content and density of the samples in these cases remained close to those values which were found earlier (see fig. 29). The exception is the index of moisture content for the first (upper) level of the model, which was reduced as a result of the drying out of the surface which was left uncovered

Table 16 *The distribution of moisture content and the final bulk weight of the skeleton*

Loadings (runs)	Scale of modelling	I level					II level					III level	
		Depth of sample retrieval		Measurements of indices			Depth of sample retrieval		Measurements of indices			Depth of sample retrieval	
		In the model (cm)	Equivalent in natural state (m)	Models		Average value	In the model (cm)	Equivalent in natural state (m)	Models		Average value	In the model (cm)	Equivalent in natural state
				18	19				18	19			
Bulk weight of the skeleton δ (g/cm^3)													
I	37.5	2.5	0.94	0.79	0.98	0.88	6.5	2.44	1.02	1.01	1.01	10.5	3
II	75	2.5	1.86	1.03	0.98	1.00	6.5	4.87	1.17	1.19	1.18	10.5	7
III	112.5	2.5	2.84	1.03	1.00	1.01	6.5	7.31	1.21	1.18	1.20	10.5	11
IV	150	2.5	3.75	1.08	1.11	1.09	6.5	9.75	1.26	1.43	1.34	10.5	15
V	187.5	2.5	4.68	1.19	1.12	1.15	6.5	12.20	1.34	1.36	1.35	10.5	1
Moisture content w of the soil by weight, as percentage													
I	37.5	2.5	0.94	26.2	25.3	25.7	6.5	2.44	26.7	26.2	26.4	10.5	3
II	75	2.5	1.86	24.8	25.4	25.6	6.5	4.87	25.8	27.0	26.4	10.5	7
III	112.5	2.5	2.84	24.7	25.0	24.9	6.5	7.31	25.4	25.5	25.5	10.5	11
IV	150	2.5	3.75	25.1	25.1	25.1	6.5	9.75	24.8	25.3	25.0	10.5	1.
V	187.5	2.5	4.68	24.7	25.1	24.9	6.5	12.20	25.2	24.7	25.0	10.5	19

Table 17 *Shear resistance τ of the soil material in non-sloping models nos. 18 and 19*

Loadings (runs)	Scales of modelling	I level					II level					III level	
		Depth of sample retrieval		τ (kg/cm^2)			Depth of sample retrieval		τ (kg/cm^2)			Depth of sample retrieval	
		In the model (cm)	Equivalent in natural state (m)	Models		Average value	In the model (cm)	Equivalent in natural state (m)	Models		Average value	In the model (cm)	Equivalent in
				18	19				18	19			
I	37.5	2.5	0.94	0.16	0.17	0.17	6.5	2.44	0.32	0.31	0.32	10.5	
II	75	2.5	1.86	0.30	0.30	0.30	6.5	4.87	0.70	0.60	0.65	10.5	
III	112.5	2.5	2.84	0.50	0.47	0.49	6.5	7.31	0.75	0.75	0.75	10.5	1
IV	150	2.5	3.75	0.50	0.50	0.50	6.5	9.75	1.07	1.05	1.06	10.5	1:
V	187.5	2.5	4.68	0.70	0.65	0.68	6.5	12.20	1.35	1.32	1.34	10.5	1

Accuracy of modelling

soil material in non-sloping models nos. 18 and 19

		IV level					V level				
asurements of ices		Depth of sample retrieval		Measurements of indices			Depth of sample retrieval		Measurements of indices		
dels	Average value	In the model (cm)	Equivalent in natural state (m)	Models		Average value	In the model (cm)	Equivalent in natural state (m)	Models		Average value
19				18	19				18	19	
13	1.12	1.12	14.5	5.45	1.25	1.35	1.30				
29	1.29	1.29	14.5	10.88	1.34	1.40	1.37				
36	1.41	1.38	14.5	16.31	1.45	1.50	1.47	18.5	20.80	1.52	–	1.52
40	1.55	1.47	14.5	21.75	1.51	1.56	1.53	18.5	27.75	–	–	–
50	1.49	1.50	14.5	27.20	1.51	1.54	1.53	18.5	34.70	1.52	1.56	1.54
2	26.3	26.3	14.5	5.45	25.9	25.6	25.7				
6	28.5	28.0	14.5	10.88	27.0	25.9	26.4				
5	26.3	26.9	14.5	16.31	27.1	26.4	26.7	18.5	20.80	28.2	26.6	27.4
8	25.6	25.7	14.5	21.75	27.3	26.2	26.7	18.5	27.75	27.8	–	27.8
8	25.1	25.4	14.5	27.20	26.4	26.0	26.2	18.5	34.70	27.5	25.7	26.6

		IV level					V level				
g/cm²)		Depth of sample retrieval		τ (kg/cm²)			Depth of sample retrieval		τ (kg/cm²)		
els	Average value	In the model (cm)	Equivalent in natural state (m)	Models		Average value	In the model (cm)	Equivalent in natural state (m)	Models		Average value
19				18	19				18	19	
	0.45	0.43	14.5	5.45	0.55	0.60	0.58				
	0.82	0.82	14.5	10.88	1.01	1.07	1.04				
	1.10	1.10	14.5	16.31	1.20	1.20	1.20	18.5	20.80	1.00	1.17	1.09
	1.27	1.30	14.5	21.75	1.35	1.30	1.33	18.5	27.75	1.02	–	1.02
	1.50	1.51	14.5	27.20	1.22	1.22	1.22	18.5	34.70	0.94	0.99	0.97

Table 18 *Shear resistance τ of soil material taken from various levels of a model (scale 187.5) and tested under a constant normal loading σ = 3.5 kg/cm²*

Levels	Depth of sample retrieval		Shear resistance τ (kg/cm²)			Moisture content w (%)			Bulk weight of the skeleton δ (g/cm³)		
	In the models (cm)	Equivalent in natural state (m)	Models		Average value	Models		Average value	Models		Average value
			18	19		18	19		18	19	
I	2.5	4.68	—	1.62	1.62	22.6[a]	24.1[a]	23.4[a]	1.08	1.13	1.10
II	6.5	12.20	1.65	1.65	1.65	26.4	25.6	26.0	1.37	1.35	1.36
III	10.5	19.70	1.45	1.52	1.48	26.0	25.2	25.6	1.44	1.49	1.46
IV	14.5	27.20	1.25	1.20	1.22	26.3	25.1	25.7	1.56	1.54	1.55
V	18.5	34.70	0.82	0.95	0.89	26.8	25.9	26.3	1.57	1.56	1.57

[a]The moisture content of the test samples from level I was partially lost on account of the drying out of the surface of the model.

Accuracy of modelling

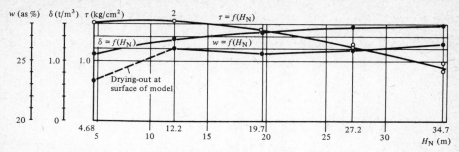

Fig. 31. Graphs showing the relationships $\tau = f(H_N)$, $\delta = f(H_N)$ and $w = f(H_N)$ for soils from model unconsolidated waste-heaps, sheared under a constant normal loading $\sigma = 3.5$ kg/cm².

during subsequent repeated sample removals.

The curve for $\tau = f(H_N)$ in the first two points, corresponding to test samples taken from marks equivalent to depths of 4.68 and 12.2 metres, is considerably elevated by comparison with the graph in fig. 30 and emerged as higher than the values corresponding to a depth of 19.7 m. In the remaining points, corresponding to depths of 27.20 and 34.70 m, the curve of the graph in fig. 31 (with duplicated measurements) duplicated the curve in fig. 30 obtained under different compressive loadings, equal to those to which the soil had been subjected during centrifuging; i.e. in the case cited for the right-hand branch of the curve the resistance of the samples of waste-heap soil which had been subjected to a greater load during centrifuging, regardless of the decrease in the compressive loading at the moment of shearing, remained the same as it had been under the action of the maximum compressive loading exerted on the soil during the centrifuging process. This is an unexpected result. It may be surmised that in the given case the structure of the soil altered with the possible squeezing out of water on to the surfaces of the elemental areas of slip formed during contraction occurring under conditions of excess pore pressure. The possibility of gross error in the determination of the points obtained in figs. 30 and 31 is excluded by the coincidence of all the determinations, repeated twice in test samples from independent models.

The data obtained during the tests on non-sloping models enabled us to establish the order of magnitude of the linear consolidation (settlement) of loose soil in model waste-heaps. In the diagram (fig. 32) the layer-by-layer values of the coefficient of linear consolidation are set out:

$$K = h_N : h'_N,$$

where h_N is the thickness of the layer in its loose state; h'_N is the thickness of the same layer after consolidation under its own weight and the weight of the overlying soil, acting during a period of time t.

For the case of the particular soils under examination a curve is presented in

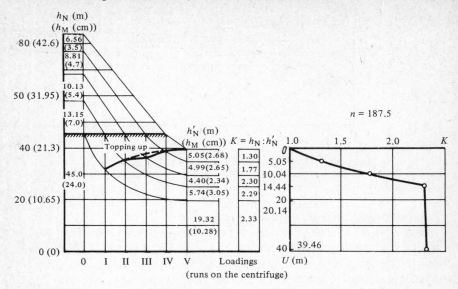

Fig. 32. Diagram illustrating the linear consolidation (settlement) of a waste-heap.

h_N (m) (and h_M (cm)) are the thickness of the layers in a loose state in nature (and in the model); h'_N (m) (and h'_M (cm)) represent the thickness of the layers after centrifuging; K ($h_N : h'_N$) is the coefficient of linear consolidation; n is the scale of modelling at the end of loading (run) V; U is the distance from the surface of a waste-heap.

fig. 32 showing the relationship between the coefficient K and the depth at which the point under consideration is situated in the waste-heap beneath its surface (U). The curve representing $K = f(U)$ reveals (in our case) an abrupt interruption at a depth of 14–15 m below which the coefficient of linear consolidation shows scarcely any further change and remains close to a magnitude of 2.30. At this point our experiments in the model testing of model waste-heaps were concluded.

As a result of the testing, in three series, of 12 sloping models with heights of 24, 30 and 36 cm, carried out to the point of collapse by landslide, and of two non-sloping models with a height of 24 cm, tested in five stages of increasing loadings (runs) up to a scale of 187.5, i.e. up to the equivalent of a natural waste-heap with a height of about 41 m (allowing for settlement of 2 metres), we were able to determine:

1. The relationship between the deformations of the models, and the density and strength of the soil materials of the models on the one hand, and the degree of loading (the number of revolutions of the model per minute) in time on the other hand.

2. The value of the coefficient of variation in the determination of the basic indices for the stability of waste-heaps, expressed in terms of H_{crit} and $\alpha = f(H)$, to within limits of 4.2%, and of the maximum errors which may arise in the

determination of these magnitudes by the centrifugal modelling method to within limits of 8.5%.

3. The ratio of the overall relative settlement $\Sigma_0^H z$ to the final height H of a waste-heap.

4. The character of the change in density (expressed in terms of the bulk weight of the skeleton) of the soil in a waste-heap through its height; when a certain limit is reached the density becomes practically constant.

5. The relationship between the shear resistance of clayey, unconsolidated waste-heap soil and the depth at which it lies in the waste-heap; it was established that up to a certain depth (corresponding, seemingly, to the spot at which the soil passes from a three-phase to a two-phase system) there occurs an increase in resistance with depth, but that beyond this a decrease in resistance occurs, in which the maximum magnitude of the normal stress σ_{max} which has been acting on the soil and of the normal load σ acting at the moment of shear is of primary significance.

6. The character of the relationship between the linear consolidation (settlement) of a waste-heap and the depth at which the soil is located in the waste-heap (calculated as the depth underground in the heap).

Modelling the natural state

The best conditions for the determination of the correspondence between models and prototypes can be represented as the parallel construction of model and prototype (in the form of an experimental waste-heap of natural magnitude) whilst observing in both, by means of geometrical similarity of shape, identical conditions in terms of the composition of the soils, their moisture content and initial density, and also conditions of equivalence of the increase in the loadings arising from the self-weight of the waste-heap. However, the organization of the construction (deposition) of such an experimental waste-heap would be too complicated and expensive. This situation was circumvented by modelling known, existing prototypes, in which capacity we adopted the actual, natural waste-heaps of overburden soil at the Novy Razdol deposits of naturally-occurring sulphur. These waste-heaps were observed and investigated by the Institutes G.I.G.Kh.S. and Ukr. N.I.I. Projekt. The Novy Razdol waste-heaps consist of quaternary sandy loams (60-80% sand) and clayey loams (25-50% clay), alluvial sands and tertiary clays, laid down either selectively or in the form of mixtures. Two types of waste-heap served as prototypes for our investigations: those consisting of a mixture of quaternary sandy loams and clayey loams and those consisting of tertiary clays.

Data concerning the prototype waste-heaps were known from the published papers of G.I.G.Kh.S. [12] and the work of L. P. Markovich [8]. In the part of interest to us there are few discrepancies between the two sources and they complement each other. Measurements of the cross-sectional profiles of both stable and sliding waste-heaps were made in large quantity by both G.I.G.Kh.S. and L. P. Markovich.

The results of the natural-state measurements of the waste-heaps by L. P. Markovich are set out in table 19. Determinations of the average angles of repose of the quaternary soils were made by him in the cases of mixtures including alluvial sands (sometimes in great quantity) and these, therefore, cannot be used for comparison with the mixtures used for our models, which did not contain these sands. The remaining measurements made by L. P. Markovich agree with the data provided by G.I.G.Kh.S.

Basing ourselves on both sources we can reckon the average indices for the actual waste-heaps at the Novy Razdol site as, characteristically: average angles of repose equal to 35° for waste-heaps of both tertiary clays and mixtures of sandy loam with clayey quaternary loams; actual limit heights for waste-heaps of mixtures of quaternary sandy loams and clayey loams will vary within the boundaries 13.7–14.7 metres, whilst that for waste-heaps of tertiary Tortonian clays is 20 metres.

The character of the collapses of the waste-heaps when they reached their critical height, at the Razdol open mines, is different for quaternary and tertiary soils. When describing the slips of the waste-heaps of quaternary soils L. P. Markovich notes that their external feature is the deformation of the lands in the form of the advance of long tongues, without any clearly delineated surface of slip.

In this case the angle of the slope of the waste-heap falls to 15–20°. The critical height of the waste-heaps in this case is very low and will vary within the limits 15–20 metres (figs. 33, 34).

In external appearance the landslips of the waste-heaps of tertiary soils differ sharply from those just described. L. P. Markovich notes that in these there is, as a rule, a well-defined back scarp (figs. 35, 36) and surface of slip. The development of the landslip occurs not gradually, but suddenly, approximating to a collapse.

The moisture content characteristics of the waste-heap soils are defined in both sources by the values set out in table 20.

Separately, L. P. Markovich presents his own determinations of the moisture content of the clayey soils in the regions of slip: $w = 29.0\%$, with plasticity limits of $w_p = 17.0\%$ and $w_L = 29.6\%$. Hence the clayey loam in the region of slip was in a state close to being fluid. In another place L. P. Markovich defines the average natural moisture content of the tertiary clays as in the limits 25–28%.

The following index of interest to us – that of the initial density of the soil at the time of its deposition in the waste-heap – was investigated by G.I.G.Kh.S. only to a limited extent. The magnitude of the coefficient of bulking was determined as equal to 1.26 for tertiary soils in an experimental waste-heap but with no indication as to the part of the waste-heap and for what duration of its lying in the waste-heap this index was determined. Markovich, on the other

Table 19 Results of measurements made by L. P. Markovich on waste-heaps at the Novy Razdol site

Character of waste-heaps	Angles of repose			Critical height				Number of determinations
	Average values of α_{rep}	Guaranteed values where $P = 0.9$		Average values of H_{crit} (m)	Guaranteed values			
		α_{rep} max.	α_{rep} min.		H_{crit} max.	H_{crit} min.		
Waste-heaps consisting of a mixture of quaternary sandy loams, clayey loams and sand	32°24'	33°34'	31°14'	13.4	14.2	12.6		43
Waste-heaps consisting of tertiary clays	34°44'	35°58'	33°30'	19.9	21.2	18.6		28

Fig. 33. Collapses by sliding of quaternary waste-heaps at the Novy Razdol open-mines, 1963 (photo by L. P. Markovich).

Fig. 34. Collapses, by landslip, of models of waste-heaps of Novy Razdol quaternary soils, 1968.

Accuracy of modelling 65

Fig. 35. The collapse, by landslip, of tertiary waste-heaps in the Novy Razdol open-mines, 1963 (photo by L. P. Markovich).

Fig. 36. The collapse, by landslip, of models of waste-heaps of Novy Razdol tertiary clays, 1968.

Table 20   *The natural moisture content, by weight, of the overburden soils of the Novy Razdol site*

	According to G.I.G.Kh.S. (per cent)			According to collations of L. P. Markovich (per cent)		
	From	To	Average value	From	To	Average value
Quaternary:						
(a) Sandy loams	11.6	23	17.6	11	23	17.3
(b) Sandy loams passing into clayey loams	–	–	–	12	25	18.5
(c) Clayey loams	18	26	20.4	10	30	23.4
Tertiary:						
(a) Clays	18	40	24.1	–	–	–
(b) Lumpy clays	–	–	–	19	40	26.8
(c) Fissured clays	–	–	–	16	35	25.0

hand, attributes great significance to the initial density of the body of the waste-heap and investigates it in detail. The actual initial density of the soil at the time of its deposition in the waste-heap, expressed in terms of the bulk weight of the skeleton δ, is determined by conversion of the data provided by Markovich and equals, in tonnes/m^3:

For quaternary sandy-loams and clayey loams: 1.26,
For tertiary clays: 1.22.

The shear-resistance of the soils is characterised in the materials provided by G.I.G.Kh.S. in the form of random figures, with no account taken of the curvature of the function $\tau = f(\sigma)$ and they are therefore of no interest to us.

L. P. Markovich compiled data-sheets to show the relationship between the shear-resistance of the soils and their degree of compaction (or looseness). Other indices of the physico-mechanical properties of the Razdol soils are set out in table 21.

The materials provided by G.I.G.Kh.S. contain information on the results of observations of an experimental waste-heap of tertiary clays constructed in 1960.

The waste-heap, in the form of a truncated cone, was built with a volume of 70 000 cubic metres on a rock foundation. It was deposited by an ESH-14/75 excavator with a capacity of up to 800 m^3/h, and the soil was levelled by bulldozer. When the waste-heap reached a height of 20 metres its collapse began in the form of the development of crevices.

The programme of investigations on the experimental waste-heap included: surveying observations, determination of the physico-mechanical properties of the waste-heap soils, a study of the structure of the waste-heap with the aid of an inspection pit, determination of the coefficient of compaction (or looseness) and investigation of the load-bearing capacity of the soils. The results of these observations were incorporated in the published papers of G.I.G.Kh.S. referred to above. They also contain a note of a 1-2° creep of the initial angle of the

Table 21 *Physico-mechanical properties of soft overburden soils from the Novy Razdol site*[a]

Soils	Plasticity (%)						Maximum molecular moisture capacity w_m (%)		Bulk weight in undisturbed body γ_b (t/m³)			Specific weight Δ (t/m³)
	Upper limit			Lower limit								
	From	To	Average value	From	To	Average value	From	To	From	To	Average value	
Sandy loams	19	26	–	13	21	–	7	12	1.94	2.15	2.06	2.67
Sandy loams passing over into clayey loams	–	–	–	–	–	–	–	–	2.04	2.25	2.10	2.6
Clayey loams	22	55	29.6	16	29	17.0	14	14.4	1.93	2.22	2.07	2.66
Lumpy clays	56	86	64.0	24	39	33.8	–	–	1.82	2.14	1.99	2.72
Fissured clays	55	89	67.5	33	46	38.4	–	–	1.81	2.15	1.99	2.86
Marl-like clays	41	81	54.0	25	32	32.6	–	–	1.88	2.27	2.09	2.70

[a]Compilation by L. P. Markovich based on works of A. S. Geraskina, K. V. Osintseva and T. V. Sirotina.

slope, revealed after a period of 16 months and observed in five profiles, and also a visual description of the structure of the experimental waste-heap, made by inspection of the walls of an inspection pit of a depth of 21.7 metres.

At the time of deposition of the experimental waste-heap the predominant dimension of the initial agglomerate lumps was 15-35 cm, but lumps measuring up to 60 cm were also encountered. In the inspection pit there were traces of abrasive action – as a consequence of local movements – on all the lumps, even the small ones. On the upper and lower surfaces of the lumps slickensides were frequently detected.

Let us cite the characterisation of the walls of the inspection pit [12]:
The character of consolidation of the soil in the body of the waste-heap can be ascertained from the walls of the inspection pit. At a depth of 2 metres the soil as a whole is only weakly consolidated: broken fragments and lumps detach themselves with ease, there is a multitude of tiny pores, and where lumps adjoin each other cavities measuring several centimetres are encountered. At a depth of 8-10 metres lumps may still be detached from the soil, but without external influence pieces of soil no longer crumble away, as was the case in the first few metres of the pit. Large cavities are no longer encountered at a depth of 8-10 metres. At depths of 17.5 and 21.5 metres the soil fragments are squeezed close to each other, and the gaps between them are filled with loose, lumpy clay.

It was originally proposed to build and test the model waste-heaps of the Razdol prototypes in two short series, in which the average characteristics of the prototypes in terms of shape, composition, moisture content, initial density and rate of waste-heap formation would be observed. However, the wide range of variations in the corresponding indices in the natural state obliged us to change over to multi-series model tests in which these indices were also varied in the models. These models were thus simultaneously included in the series of fundamental investigations which served to determine the degree of influence of various different factors on the stability of unconsolidated waste-heaps.

The models of the Razdol waste-heaps, of both tertiary and quaternary soils, were made to precisely the same pattern as the models constructed from Kerchi tertiary clays, i.e. with an initial angle of slope of $36°$. The angle of repose of the Razdol waste-heap soils is in both cases a mere $1°$ less than the initial angle of slope in the models. During centrifuging there occurs not only in the first but also in all the subsequent runs (loadings) a certain contraction over the whole external contour of the model with a gradually attenuated flattening of the angle of the slope. During the testing of the previous models it was discovered that, as a rule, collapses occur at the resultant angles $1-2°$ less than the initial angle of slope, i.e. within the limits of the average measured angles of repose in the Razdol waste-heaps.

The compositions of the soil materials of the models correspond approximately to the average soil compositions of the two prototype waste-heaps: mixtures of

Accuracy of modelling

quaternary sandy loams and clayey loams from those areas of the site that have no sandy alluvial deposits (in the first case) and waste-heaps of tertiary clays (in the second case). For brevity we shall call the first group of waste-heaps and their models quaternary (sandy loams + clayey loams) and the second group tertiary (clays).

The composition of the materials of the model quaternary waste-heaps include sandy loams, sandy loams passing into clayey loams and clayey loams, in the proportions of 5 : 2 : 1.

The materials comprising the model tertiary waste-heaps include Tortonian clays, both grey and bluish-grey, dense, micaceous, sandy and greasy, lumpy and schisted, with interlayers of bentonitic clay, sometimes marly; i.e. all the various kinds of tertiary clays encountered in the composition of the overburden of the site, taken in an amount proportional to their composition in the core-samples of the tertiary stratum.

The starting material for the construction of the models was selected from the northern section of the Razdol site in the form of core-samples, preserved in paraffin, of three specially assigned bore-holes. The physico-mechanical properties of the starting materials for the models are set out in table 6 and in figs. 20 and 21.

The moisture content of the soil materials of the models will vary within the following limits, in percentage;

For sandy loams + clayey loams: 20-22.5-25,
For tertiary clays: 28-31-32.

The initial density of the soil masses (expressed in terms of the bulk weight of the skeleton) at the time of deposition in the models lies within the limits, in tonnes/m^3:

For sandy loams + clayey loams: 0.91-1.22-1.54,
For tertiary clays: 0.75-1.00-1.21.

The duration of the loadings (runs) on the models was taken as

$$t_I = 44'48'', \quad t_{II} = 14'48'' \quad \text{and} \quad t_{III} = 6'00'',$$

which corresponds to delivery rates of the waste-heap forming equipment or excavators working on the waste-heap, in terms of m^3/h of loose material, equivalent to:

$$Q_I \approx 800, \quad Q_{II} \approx 2400 \quad \text{and} \quad Q_{III} \approx 5950.$$

Descriptions and results of the tests on the models of the Razdol waste-heaps are set out below, as part of the series of tests to establish the general laws governing the stability of an unconsolidated waste-heap based on concrete factors. What now interests us is only the degree to which they correspond to the natural state.

By processing the results using the method of mathematical statistics simple empirical formulae were obtained for the relationship between the critical height of a waste-heap H_{crit} and the moisture content of the soil w, its initial density in the waste-heap δ and the delivery rate of the waste-heap forming machinery Q. For the Razdol waste-heaps these formulae are:

For quaternary soils, in metres

$$H_{crit} = 103.22 - 3.41w - 7.31\delta - 0.01663Q + 0.001978Q + 0.00052wQ.$$

For tertiary soils, in metres

$$H_{crit} = 177.27 - 4.83w - 9.08\delta - 0.031Q + 0.001wQ.$$

In these we substitute the values, corresponding to the natural state conditions, of the variables Q, δ and w.

The waste-heap formation capacity using the ESH-14/17 excavator, employed on the waste-heap, was precisely $Q = 800$ m^3/h of loose material.

The average initial bulk weight of the skeleton of the soil mass in the waste-heaps was, (according to L. P. Markovich), in tonnes/m^3

For the quaternary waste-heaps: 1.26,
For the tertiary waste-heaps: 1.22.

The limits of the moisture content of the soil in the natural state (see table 20) vary over a wide range and their average values are defined differently by each investigator. Obviously, it is impossible to insert them directly into the calculation as a starting index. Hence, using the values of the critical heights of the waste-heaps, H_{crit}, in metres, established for the natural state, i.e.

For the quaternary waste-heaps: 13.7–14.7,
For the tertiary waste-heaps: 20,

and the derived empirical relationships, we solve the inverse problem by sequential approximation: we find the 'critical moisture content' of the soil materials in the waste-heaps, at which the failure of the waste-heaps begins in the natural state.

For the quaternary waste-heaps

$$w_{crit} = \frac{103.22 - 7.31\delta - 0.01663Q + 0.001978Q - H_{crit}}{3.41 - 0.00052Q} = 23\text{--}22.7\%.$$

For the tertiary waste-heaps

$$w_{crit} = \frac{176.8 - 9.1\delta - 0.031Q - H_{crit}}{4.83 - 0.001Q} = 27.6\%.$$

In the first case we have almost total conformity of the derived 'critical moisture content' (23–22.7%) with the average moisture content of the clayey loams (23.4%) and the maximum moisture content of the sandy loams (23%), as observed in the natural state by L. P. Markovich and the collaborators at G.I.G.Kh.S.

Accuracy of modelling

In the second case the moisture content $w_{crit} = 27.6\%$ is only 0.4% lower than the upper limit of the natural moisture content of clays, as observed by L. P. Markovich in the natural state (28%) and 0.8% higher than the average values of the natural moisture content of lumpy clays (26.8%), as determined by the predecessors of L. P. Markovich. The critical moisture content obtained by calculation does not exceed the limits of the moisture content of the clays observed in the natural state (see table 20) and differs by only 1.4% from the moisture content of the tertiary clays in the region of the slipping waste-heaps, as observed by L. P. Markovich.

The character of the failures of the models of tertiary and quaternary waste-heaps was sharply differentiated. The descriptions given by L. P. Markovich of the character of the deformations of both types of waste-heap are also acceptable to describe the kinds of deformations of the models of tertiary and quaternary waste-heaps.

The visual description given by G.I.G.Kh.S. of the cross-section of the waste-heap in the inspection pit also accords with the cross-sections of the models at the scale of the latter stage of loading (during the final run).

Thus, from a comparison of the natural state with the models the correspondence of the conditions of similarity is confirmed and its degree of accuracy can be considered established by the results of the comparative investigations of models of different scales and of the order of 7–8%.

3

The relationship between the stability of unconsolidated waste-heaps and the moisture content of the soils, their initial density in the waste-heap and the rate of formation of the waste-heap

Five basic factors determining the conditions of the stability of unconsolidated waste-heaps on a firm, horizontal foundation

We express the stability of unconsolidated waste-heaps in terms of the property of the critical height of the waste-heap H_{crit} at a known value of the angle of repose α_{rep} and by the relationship $\alpha_{lim} = f(H)$, when $H > H_{crit}$. This latter relationship can have two values. The first is when a waste-heap with a height $H > H_{crit}$ is deposited with a slope which has a resulting angle α'_{lim} smaller than the angle of repose α_{rep}. The second is when the slumping of the slope to reach its limit (critical) angle α''_{lim} occurs after the formation of a continuous surface of slip in the body of the waste-heap as a result of the exceeding of the limit (critical) height H_{crit}.

Excluding the effect of a weak or inclined foundation, i.e. considering only the cases of the location of waste-heaps on a firm, horizontal foundation, the conditions of the stability of unconsolidated waste-heaps are determined (in the main) by five factors: the composition of the soil mass of the waste-heap, its moisture content, initial density when deposited in the waste-heap, the rate of deposition (the rate of waste-heap formation) and the method of depositing the soil in the waste-heap.

On deeper analysis, it appears that each of the main factors listed can be regarded as dependent on secondary factors which enter into our determinations only in their overall manifestations.

For example, the duration of the time during which a waste-heap that has been deposited but not yet consolidated is left to stand shows itself in the extent of the continuing binding together of the soil aggregates in the waste-heap, and is reflected in the development of pore pressure, and also in its dissipation. In this way the duration of interruptions in the process of waste-heap formation, or the time after its completion, influences the degree of consolidation of the soil masses in the waste-heap and, consequently, the conditions of the stability of the waste-heap also. In our solutions time enters our considerations only in terms of the studied factor of the rate of increase of the loadings before and after the beginning of the failure of the waste-heap.

However, the centrifugal modelling method does not exclude the possibility of special investigations into the narrow question of the influence of the period a deposited waste-heap is allowed to stand on changes in the physico-mechanical properties of its component soil masses, and in particular on the conditions of its stability in time after completion of the deposition.

At present we do not command the means to compare simultaneously the influence of all the five main factors elucidated; we shall therefore resolve the stated problem in groups, which will simultaneously include not more than three variables.

To the first group we shall assign the relationships $H_{crit} = f(w, \delta, Q)$.

The second and third groups present themselves as offshoots from the basic model with changes in the composition of the soil material, accountable in terms of changes in the correlation of inert/sandy and active/clayey components of the mixture – in the former case. In the latter case we model changes in the methods of waste-heap formation, which occur in three variations: the deposition of large fragments in horizontal layers in bulldozed waste-heaps, bridge deposition with several trickle-streams on the bridge cantilever, and deposition by waste-heap former with a single stream in thin, parallel slanting layers.

In the beginning, experiments of the first group were carried out with investigations on 36 pairs of models, constructed from three types of soil. As has already been indicated, in the construction of the models we used as starting soil materials the quaternary (sandy loams + clayey loams) and tertiary (Tortonian clays) soils studied earlier, from the overburden of the Novy Razdol natural sulphur deposits, and also Kerchi tertiary (heavy) clays, which had been used in the first stage of our investigations. In addition, one pair of models was prepared from pure Dnieper sands, which later served as an inert addition when testing models constructed from mixed soils.

The physico-mechanical characteristics of all the original starting soil materials have been presented earlier. The strength parameters were determined as applicable to different states of the soil materials, corresponding to those conditions in which they found themselves in the models.

Plan of the experiment

The plan of the experiment to determine the relationship $H_{crit} = f(w, \delta, Q)$, with the given types of soil and methods of deposition, must satisfy the conditions which afford the possibility of investigating the coupled relationships

$H_{\text{crit}} = f(w)$, $H_{\text{crit}} = f(\delta)$ and $H_{\text{crit}} = f(Q)$, and the possibility of the mathematical expression of the overall relationship $H_{\text{crit}} = f(w, \delta, Q)$ by one of the statistical methods.

The first condition is satisfied by the disposition of the models in three mutually intersecting directions: the horizontal w, the vertical Q, and the oblique (transverse) δ, with three coupled models in each direction, i.e. seven pairs of models in all (one pair is common to all three directions).

To satisfy the second condition in the simplest case one requires two points (two pairs of models) for each variable in all possible variations of combination with the two other variables (see table 26) to make possible the construction of an equation of the type:

$$H_{\text{crit}} = b_0 + b_1 w + b_2 \delta + b_3 Q + b_{12} w\delta + b_{13} wQ + b_{23} \delta Q + b_{123} w\delta Q,$$

where $b_0, b_1, b_2, \ldots, b_{123}$ are certain coefficients which can be determined by mathematical, statistical methods.

For this eight pairs of models are required, only partly conforming with the models which satisfy the first condition.

Systematising all the models in an overall plan of experimentation, we define the position of the given models in tables 22 and 23, constructed in accordance with a single principle and including eleven pairs of models for each of the types of soil materials from the Novy Razdol open mines which we investigated. In table 24 the plan is presented of the abbreviated experiment with Kerchi clays, which provided for the finding of only three pairs of relationships.

The extreme values of the variable w are fixed in accordance with the conditions: firstly, of closeness to the values in the natural state and, secondly, the possibility of handling the soil materials for the construction of models, since when beginning with some optimum moisture content a deconsolidated clayey material may be transformed into a sticky, mud-like mass, unsuitable for the construction of models.

The limit values of the moisture content w in our experiments were taken as: for quaternary soils from Novy Razdol (sandy loams plus clayey loams) in the main group of tests - from 22.4 to ~25% and additionally at 20%; for tertiary (Tortonian) clays from Novy Razdol in the main group of tests - at 28.1 and 30.8% and additionally at 32%; for Kerchi tertiary clays in the main group of tests - at 28.0 and 31.0% and additionally at 25%.

The limit values of the initial density of the soil material δ in the model waste-heaps were fixed on the basis of the following possibilities: on the one hand stands the loose deposition of the soil 'from the shovel' (with only light levelling by the mould) and, on the other hand maximum dense deposition of the soil in the model, with ramming to a constant volume.

However, in the latter case it is not possible to achieve a uniform density of the body of the model, and this leads to significant scattering of the results of the experiment, sometimes beyond the boundaries of the extreme values. It

Table 22   *Plan of the experiment with model waste-heaps of Novy Razdol quaternary sandy loams and clayey loams to determine the degree of the influence of the moisture content w, the initial density δ and rate of waste-heap formation Q on the critical height of the waste-heaps H_{crit}*

Rate of formation Q (m³/h)	Initial density δ (t/m³)	Series	Moisture content w (%) ~25		~22.5		~20.3
			Initial density δ (t/m³) 1.08	~1.53	~0.91	~1.53	0.98
			C'	C''	D'	D''	
800 (t_M^I = 44'48")	0.99	A'	⊠ 22–23 H_{crit} = 8.7		24–25 ⊙ H_{crit} = 16.3 ● 28–29 H_{crit} = 14.3		△ 20–21 H_{crit} = 30.7
	1.53	A''		■ 64–65 H_{crit} = 4.5		26–27 H_{crit} = 14.3	
2400 (t_M^{II} = 14'48")	1.00	B'	⊠ 58–59 H_{crit} = 5.6		⟡ 56–57 H_{crit} = 12.0		*Series E (oblique)*
	1.53	B''		⊠ 62–63 H_{crit} = 4.3		◆ 60–61 H_{crit} = 10.5	
5950 (t_M^{III} = 6'00")	1.00				☼		

Note: The density of models 28–29 was δ = 1.22 t/m³

is obviously expedient to reject this third, intermediate point in the determination of $H_{crit} = f(\delta)$ and thus restrict the number of tests in the overall plan to ten pairs of models for each type of waste-heap soil.

In the experiments carried out the limit values of the initial consolidation δ were equal to, in g/cm³:

For models of the Novy Razdol quaternary waste-heaps (sandy loams + clayey loams): 0.91–1.54.

For models of the Novy Razdol tertiary waste-heaps (clays): 0.75–1.21.

For models of the Kerchi tertiary waste-heaps (clays): 0.67–1.43.

The grounds for adopting the values Q_I = 800 m³/h and Q_{II} = 2400 m³/h for the rate of loading the waste-heaps or the rate of waste-heap formation were set out earlier. In the case of the third value of Q_{III} = 5950 m³/h the rate of loading the model, defined by the limit speeds of acceleration of the centrifugal machine, is so great that the deformations of the model lag behind the loading.

We can recommend determination of the third point of the curve $H_{crit} = f(Q)$ not at the maximum, but at the minimum realistically possible rate of waste-heap formation.

Table 23 Plan of the experiment with model waste-heaps of Novy Razdol tertiary clays to determine the degree of influence of the moisture content w, initial density δ and rate of waste-heap formation Q on the critical height of the waste-heaps H_{crit}

Rate of formation Q (m³/h)	Moisture content w (%)		~32		~31			~28	
	Initial density δ (t/m³)	Initial density δ (t/m³) \ Series	0.75	1.21	0.75	1.21		0.75	1.21
800 ($t_M^I = 44'48''$)		0.75	A' ⊠ 38–39 $H_{crit}=15.2$		C' ⊙ 36–37 $H_{crit}=21.5$	C'' 42–43 $H_{crit}=19.5$ ●	Series E (oblique) 40–41 $H_{crit}=15.0$	D' △ 34–35 $H_{crit}=30.1$	D'' ▲ 54–55 $H_{crit}=26.5$
		1.21	A''				52–53 $H_{crit}=27.3$ △		
2400 ($t_M^{II} = 14'48''$)		0.75	B'		⊕ 44–45 $H_{crit}=21.8$	48–49 $H_{crit}=18.2$ ◆			▲ 50–51 $H_{crit}=24.3$
		1.21	B''						
5950 ($t_M^{III} = 6'00''$)	~0.75				☼ 46–47 $H_{crit}=22.5$				

Note: The density of models 42–43 was $\delta = 0.98$ t/m³.

Table 24 *Plan of the abbreviated experiment model waste-heaps of Kerchi tertiary clays to determine the degree of the influence of the moisture content w, initial density δ and rate of waste-heap formation Q on the critical height of the waste-heaps H_{crit}*

Rate of formation Q (m³/h)	Initial density δ (t/m³)	Moisture content w (%) Initial density δ (t/m³) Series	~31 0.67 C'	1.43 C"	~28 0.67 D'	1.43 D"	~25 0.67
800 (t_M^I = 44'48")	0.67	A'	⊠ 68–69 H_{crit} = 13.7		⊙ 66–67 H_{crit} = 20	● 74–75 H_{crit} = 21.2 ● 70–71 H_{crit} = 18.4	△ 72–73 H_{crit} = 24.9
	1.43	A"		■			
2400 (t_M^{II} = 14'48")	0.67	B'	⊠		◇ 76–77 H_{crit} = 16.6		Series E (oblique)
	1.43	B"		✖			
5950 (t_M^{III} = 6'00")	0.67				✦ 78–79 H_{crit} = 16.6		

Notes: 1. The density of models 74–75 was δ = 1.06 t/m³.
2. Supplementary 'rain' models 80–81 were with the conditions whereby: δ = 0.67 t/m³. Q = 800 m³/h; $w = w' + w''$ = 28%, where w' = 20% – uniformly distributed moisture content; w'' = 8% – surface wetting of the material immediately before deposition in the model.

In the models tested according to the basic plan of the experiment (to find H_{crit}) the curve for the relationship $\alpha''_{lim} = f(H)$ is also determined. In order to obtain the additional points of this curve, after the collapse of each model, the test is extended for just one further loading (run) with prior topping-up of the collapsed model back to its original height.

In order to construct the curve of the relationship $\alpha'_{lim} = f(H)$ on sections corresponding to heights H which significantly exceed the critical height of the waste-heap it is necessary to test an additional series, consisting of not less than three pairs of models, including two with initial heights greater than the initial height of the models of the main experiment, and with initial angles of slope less than the angle of repose.

Determinations of the critical height H_{crit}, and of the relationship $\alpha_{lim} = f(H)$

The information on the emergence and course of the deformations of models, collated in tables (in accordance with the example of appendix 1) which represent a conversion, and the citing of observations of models as equivalents of waste-heaps in the natural state, serve as the basic numerical material for the construction of graphs of the relationship between the critical angles of the slopes and the height of the waste-heaps in different conditions of moisture content, initial density of the soils in the waste-heap and the rate of waste-heap formation.

The models are tested in series which correspond to the investigations of the influence of each distinct variable.

In accordance with the experimental plan graphs are constructed of the relationship $\alpha_{lim} = f(H)$ and the critical height H_{crit} is also determined by series of tests:

Series A' With variable moisture content, loose and slow deposition.
A'' With variable moisture content, dense and slow deposition.
B' With variable moisture content, loose and rapid deposition.
B'' With variable moisture content, dense and rapid deposition.
C' With variable rate of deposition, high moisture content and loose deposition.
C'' With variable rate of deposition, high moisture content and dense deposition.
D' With variable rate of deposition, low moisture content and dense deposition.
D'' With variable rate of deposition, low moisture content and dense deposition.
E With variable density, but constant moisture content and rate of deposition.
F With variable initial angles of slope α and constant moisture content, density and rate of deposition.

The plans of the experiments are set out in tables 22, 23 and 24, and summary graphs of the relationship $\alpha_{lim} = f(H)$ and of the determination of H_{crit} for Novy Razdol quaternary waste-heaps (sandy loams plus clayey loams) are given in fig. 37, and those for Novy Razdol tertiary waste-heaps (Tortonian clays) in fig. 38.

In figs 39, 40 and 41 examples are shown of the construction of graphs, by series, of the tests on waste-heaps of Kerchi tertiary clays.

The nature of the failure of waste-heaps when their critical height is exceeded, in relation to the lithologic composition of the soils deposited in the waste-heap, is diverse. When the failure or collapse of waste-heaps of clayey soils occurs, one (or more rarely two or three) continuous and complete surface of slip develops, with the formation of which the resulting critical angles of the slopes of the waste-heap α''_{lim}, given the same height of the waste-heap, are sharply reduced and are determined not by the strength of the waste-heap soil in its mass but by

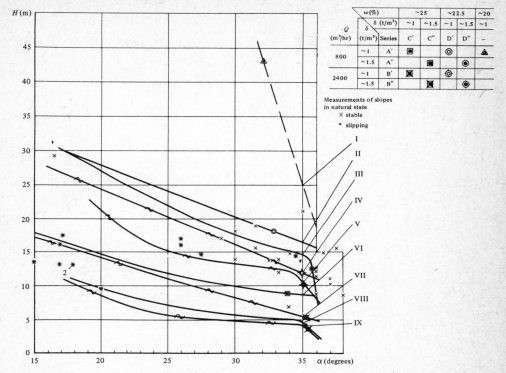

Fig. 37. The relationship $\alpha_{lim} = f(H)$. Summary. Sandy loams plus clayey loams from Novy Razdol:

I: $\dfrac{H_{crit} = 30.7 \text{ m}}{\text{Models 20–21}}$; II: $\dfrac{H_{crit} = 16.3 \text{ m}}{\text{Models 24–25}}$; III: $\dfrac{H_{crit} = 14.3 \text{ m}}{\text{Models 26–27}}$;

IV: $\dfrac{H_{crit} = 12.0 \text{ m}}{\text{Models 56–57}}$; V: $\dfrac{H_{crit} = 10.5 \text{ m}}{\text{Models 60–61}}$; VI: $\dfrac{H_{crit} = 8.7 \text{ m}}{\text{Models 22–23}}$;

VII: $\dfrac{H_{crit} = 5.6 \text{ m}}{\text{Models 58–59}}$; VIII: $\dfrac{H_{crit} = 4.5 \text{ m}}{\text{Models 64–65}}$; IX: $\dfrac{H_{crit} = 4.3 \text{ m}}{\text{Models 62–63}}$.

the shear resistance along the entire continuous surface of weakening that has arisen in the form of a slickenside in the body of the waste-heap. Hence, for waste-heaps of clayey soils two states are distinguished with differing boundary norms of stability: before and after the development of continuous surfaces of slip. In the graphs this is reflected in two separate curves (see figs. 38, 39, 40 and 41) as follows:

$\alpha'_{lim} = f(H)$, corresponding to the relationship before surfaces of slip develop.
$\alpha''_{lim} = f(H)$, corresponding to the relationship after the development of a surface of slip.

The critical height of the waste-heap is found by reference to the graph for

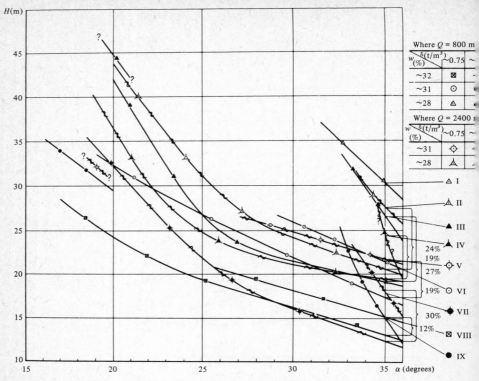

Fig. 38. The relationship $\alpha_{lim} = f(H)$. Summary. Novy Razdol tertiary clays:

I: $\dfrac{H_{crit} = 30.1 \text{ m}}{\text{Models 34-35}}$; II: $\dfrac{H_{crit} = 27.3 \text{ m}}{\text{Models 52-53}}$; III: $\dfrac{H_{crit} = 26.5 \text{ m}}{\text{Models 54-55}}$;

IV: $\dfrac{H_{crit} = 24.3 \text{ m}}{\text{Models 50-51}}$; V: $\dfrac{H_{crit} = 21.8 \text{ m}}{\text{Models 44-45}}$; VI: $\dfrac{H_{crit} = 21.5 \text{ m}}{\text{Models 36-37}}$;

VII: $\dfrac{H_{crit} = 18.2 \text{ m}}{\text{Models 48-49}}$; VIII: $\dfrac{H_{crit} = 15.2 \text{ m}}{\text{Models 38-39}}$; IX: $\dfrac{H_{crit} = 15.0 \text{ m}}{\text{Models 40-41}}$.

$\alpha_{lim} = f(H)$ at the point of intersection of the curve with the ordinate $\alpha = \alpha_{rep}$. In our cases this ordinate was $\alpha = 35°$.

The distance between the two curves where the values of α are identical defines the magnitude of the reduction in the critical height of the stable waste-heap after the development of surfaces of slip and henceforward will be denoted as h_{red}. The extent of the reductions in the critical height of the waste-heaps in our cases, after the development of surfaces of slip, varied from 11.9 to 30.7%, with an average of about 20%, of the original height of the waste-heap.

From the moment of the emergence of a protrusion beyond the contours of the slope, deposited at the angle of repose, and the development of local

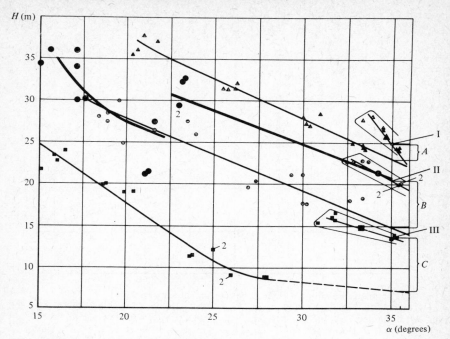

Fig. 39. The relationship $\alpha_{\lim} = f(H)$. Series A'. Kerchi tertiary clays.
I: models 72–73 ($H_{crit} = 24.9$ m; $w = 25.3\%$; $\delta = 0.705$ t/m^3; $t = 44'48''$; $Q = 800$ m^3/h); II: models 66–67 ($H_{crit} = 20.5$ m; $w_{init} = 28.4\%$; $\delta_{init} = 0.672$ t/m^3; $t = 44'48''$; $Q = 800$ m^3/h); III: models 68–69 ($H_{crit} = 13.7$ m; $w_{init} = 31.2\%$; $\delta_{init} = 0.668$ t/m^3; $t = 44'48''$; $Q = 800$ m^3/h); A = reduction of 7.2%; B = reduction of 27.3%; C = reduction of 47.5%.

slickensides in the body of the waste-heap until the moment when they merge into a continuous surface of slip through the whole of the waste-heap a transitional state exists. During this period the height of the waste-heap remains close to the critical height. An intensive increase in the size of this protrusion serves to signify a state of imminent slip.

Protrusion of the slope also occurs in waste-heaps composed of sandy loams, but waste-heaps of sandy loams which have exceeded their critical height fail and collapse in the form of multi-step (jerky) movements without clearly defined slickensides. In some cases the steps succeed each other so rapidly and frequently that the collapse takes on the character of an outward flow with the formation, in the lower part, of a bulging convex contour (figs. 33, 34, 37).

The conditions established for the series F models, which were used for the determination of $\alpha'_{\lim} = f(H)$ before the onset of collapse of high waste-heaps with reverse angles of slope (in comparison with the angle of repose) are set out in table 25, whilst an example of the results obtained from the tests of one of these series is indicated by the heavy line in fig. 39 (Kerchi tertiary clays). Standing outside of the overall plan are the tests of models 82 and 83, models of

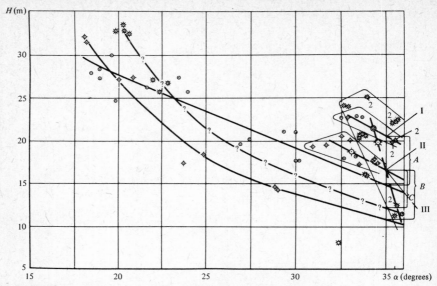

Fig. 40. The relationship $\alpha_{lim} = f(H)$. Series D'. Kerchi tertiary clays.
I: models 66–67 (H_{crit} = 20.0 m, w_{init} = 28.4%; δ_{init} = 0.672 t/m³; t = 44'48"; Q = 800 m³/h); II: models 76–77 (H_{crit} = 16.6 m; w_{init} = 28.4%; δ_{init} = 0.671 t/m³; t = 14'48"; Q = 2400 m³/h); III: models 78–79 (H_{crit} = 16.6 m; w_{init} = 28.3%; δ_{init} = 0.672 t/m³; t = 6'00"; $Q \approx$ 6000 m³/h); A = reduction of 27.3%; B = reduction of 34.3%; C = reduction of 26.5%.

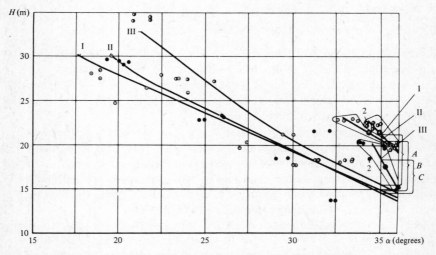

Fig. 41. The relationship $\alpha_{lim} = f(H)$. Series E (oblique). Kerchi tertiary clays.
I: models 74–75 (H_{crit} = 21.2 m; w_{init} = 28.3%; δ_{init} = 1.055 t/m³; t = 44'48"; Q = 800 m³/h); II: models 66–67 (H_{crit} = 20.0 m; w_{init} = 28.4%; δ_{init} = 0.672 t/m³; t = 44'48"; Q = 800 m³/h); III: models 70–71 (H_{crit} = 18.4 m; w_{init} = 28.4%; δ_{init} = 1.433 t/m³; t = 44'48"; Q = 800 m³/h); A = reduction of 26.9%; B = reduction of 27.3%; C = reduction of 20.6%.

Table 25 Original starting parameters of series F models

Type of waste-heap	Initial parameters of the models										
	Basic						Supplementary				
	Model numbers	w (%)	δ_{init} (g/cm^3)	t	H_M (cm)	α	Model numbers	w (%)	δ_{init} (g/cm^3)	t	H_M (cm)	α
Quaternary sandy loams plus clayey loams (Novy Razdol)	22–45	22.4	0.91	44'48"	24	36°00'	94	22.4	0.91	44'48"	36	28°30'
Tertiary Tortonian clays (Novy Razdol)	36–37	30.9	0.75	44'48"	24	36°00'	95	30.5	0.75	44'48"	36	28°30'
Tertiary clays (Kerchi)	66–67	28.4	0.67	44'48"	24	36°00"	96–97	28.7	0.67	44'48"	24	25°00'

Note: The duration of the runs $\tau = 44'48''$ corresponds to a rate of deposition $Q = 800$ m^3/h.

sandy embankments formed from loosely deposited (where $\delta = 1.52$ t/m^3) Dnieper sand (see table 6, fig. 19) which later served as the inert additive in investigations of the stability of mixed waste-heaps (clay plus sand). For these the earlier original shape of the model was retained, but with a slope at an angle $\alpha = 32°$, which exceeded by one degree the angle of repose $\alpha_{rep} = 31°$. The initial height of the model was 24 cm, and the initial moisture content of the sand $w_{init} = 13.6\%$ (capillaric wetting). The conditions of the loading of the models corresponded to a rate of deposition of the waste-heap $Q = 800$ m^3/h.

Acceleration of the models was carried out up to 302 revs/min and the scale of modelling to $n = 247$. At this scale the deformations of the models took the form of only insignificant settlement of the upper surface area (by 2.5 cm).

The main mass of water was squeezed out through the sandy base with no signs of any flowing or spreading of the sands in the slopes. The final moisture content of the sands turned out to be equal to, in percentage:

At the surface: 2.64.
At a depth of 8 cm: 3.05.
At a depth of 16 cm: 3.24.
At a depth of 22 cm: 3.41.

Thus as a result of centrifuging, the moisture content of the sands reached a norm close to the maximum molecular moisture capacity of $w_{mol} = 3.04\%$.

Investigation of the influence of the moisture content and initial density of soils and of the rate of waste-heap formation on the stability of waste-heaps

The results of our investigations set out in figs. 37–41 were obtained in conditions of variable moisture content w, initial density δ and rate of deposition of the waste-heaps Q, all relatively close to the limits observed in the natural state. Within these limits they show a reduction in the critical height of the waste-heaps with an increase in the values of all three variables, although the character and extent of the influence of each are different.

The influence of the moisture content of waste-heap soils

The indices of the stability of waste-heaps H_{crit} and $\alpha_{lim} = f(H)$ which we adopted for waste-heap soils containing clayey and dust-like components are very sensitive to changes in moisture content (figs. 42, 43, 44). This relationship is two-staged: a direct relationship, determined solely by variations in moisture content, where the soil composition of the skeleton is definitely preserved, and an oblique relationship, involving a sensitive reaction to the changes in the lithologic composition, which determines the water characteristics of the soil.

The relative reduction in the critical height of a waste-heap in relation to moisture content (within the limits which are normal in natural conditions) given

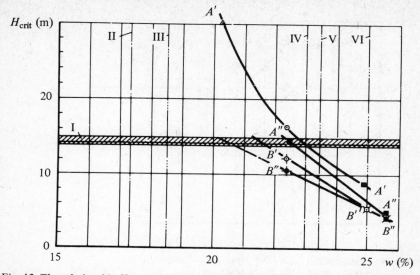

Fig. 42. The relationship $H_{crit} = f(w)$. Novy Razdol sandy loams plus clayey loams.
A', the series of loosely deposited models ($\delta_{init} \approx 0.95$ t/m³) at a rate of deposition $Q \approx 800$ m³/h ($t = 44'48''$); A'', the same, but with dense deposition ($\delta_{init} = 1.53$ t/m³) at a rate of deposition $Q = 800$ m³/h ($t = 44'48''$); B', the same, but with loose deposition ($\delta = 0.95$ t/m³) at a rate of deposition of 2400 m³/h ($t = 14'48''$); B'', the same, but with dense deposition ($\delta = 1.53$ t/m³) at a rate of deposition of ~2400 m³/h ($t = 14'48''$); I, average critical heights of the waste-heaps in the natural state; II, average moisture content of the sandy loams; III, average moisture content of the sandy loams passing into clayey loams; IV, maximum moisture content of the sandy loams; V, average moisture content of the clayey loams; VI, maximum moisture content of soils intermediate between sandy loams and clayey loams.

an increase in moisture content of 1 per cent will vary as follows, in percentage:
  In Novy Razdol quaternary waste-heaps: 25–40 (absolute reduction up to 8 m).
  In Novy Razdol tertiary waste-heaps: 12–16 (absolute reduction up to 4 m)
  In Kerchi clay waste-heaps: 10–14 (absolute reduction up to 2 m).
  The limits are very great and are sensitive to the lithologic composition of the soil masses of the waste-heaps.
  In our cases the waste-heaps masses composed of sandy loam and clayey loams show a much greater reaction to the additional wetting, and in the models and in the natural state alike retain their stability much worse than tertiary clays. Evidently, in these cases we see the effect of the large differences in the values of the limit of molecular moisture capacity, which in the case of clays exceeds by many times the water-retaining capacity of sandy loams. However, it would be an absurdity to consider a superiority in hydrophilic nature to be always a positive factor in relation to the stability and shear-resistance of soil material. The influence of an inert addition of sands to clayey waste-heap masses likewise does not always produce the positive effect of an increase in the stability of a waste-heap, as is shown later.

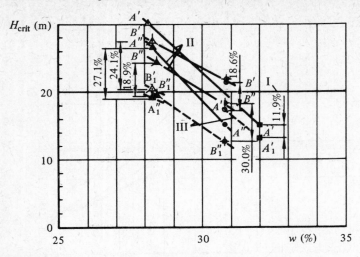

Fig. 43. The relationship $H_{crit} = f(w)$. Novy Razdol tertiary clays.
A', the series of models with loose deposition (δ_{init} = 0.75 t/m³) at a rate of deposition of ~ 800 m³/h (t = 44'48''); A'', the same, but with dense deposition (δ_{init} = 1.21 t/m³) at a rate of deposition of ~ 800 m³/h (t = 44'48''); B', the same, but with loose deposition (δ_{init} = 0.75 t/m³) at a rate of deposition of 2400 m³/h (t = 14'48''); B'', the same, but with dense deposition (δ_{init} = 1.22 t/m³) at a rate of deposition of 2400 m³/h (t = 14'48''); I, average critical height of waste-heaps in natural state; II, (solid line) before formation of surface of slip; III, (dotted line) after formation of surface of slip.

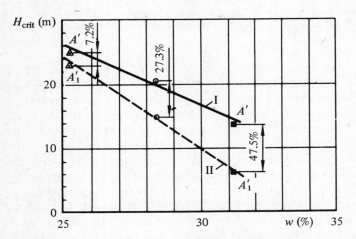

Fig. 44. The relationship $H_{crit} f(w)$, Kerchi tertiary clays. Loose deposition ($\delta_{init} \approx 0.68$ t/m³). Rate of waste-heap formation $Q \approx 800$ m³/h (t = 44'48'').
I, before development of surface of slip; II, after development of surface of slip.

The influence of the initial density of the soils in a waste-heap

Investigations of the extent to which the stability of waste-heaps is dependent on the initial density of the soil masses (in the waste-heaps) (figs. 45, 46, 47) were carried out by us in conditions whereby their moisture contents were kept constant, i.e. where the possibility of rapid evacuation of excess water from the body of the waste-heap is excluded.

The use of a drain in the body of the waste-heap alters the essence of the problem: it opens a face outlet for the expelled water, reduces the degree of wetting and promotes a sharp increase in stability.

We express the density of the soil mass in terms of the bulk weight of the skeleton δ, which varied in our model waste-heaps from 0.7 to 1.6 t/m^3, limits which were fixed on the one hand by the possibility of loose deposition 'from the shovel' and, on the other hand, by the maximum initial consolidation possible by means of ramming, in layers, to achieve a constant volume.

Forced compaction squeezes out a significant part of the gaseous component (third phase) of the soil and forces the hard aggregates closer together. At the same time the cohesion between the aggregates (adhesion and interlocking) is increased. The soil aggregates cease to roll off down the slope and the slope itself acquires the ability to retain a vertical position up to a certain height. However, in spite of this, the critical height of a waste-heap, deposited with compaction at the angle of repose α_{rep}, frequently turns out to be less than the critical height of a loosely deposited waste-heap.

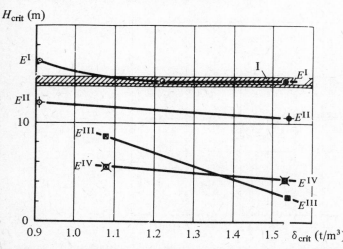

Fig. 45. The relationship $H_{crit} = f(\delta_{init})$. Sandy loams plus clayey loams from Novy Razdol.
I, average critical values of the height of the waste-heaps in the natural state; E^I and E^{II} series of models with a rate of deposition of $Q = 800$ m^3/h ($t = 44'48''$) and where $w_{init} \approx 22.4\%$ and $w_{init} \approx 25\%$ respectively; E^{III} and E^{IV} are series of models with a rate of deposition of $Q = 2400$ m^3/h ($t = 14'48''$) and where $w_{init} \approx 22.4\%$ and $w_{init} \approx 25\%$ respectively.

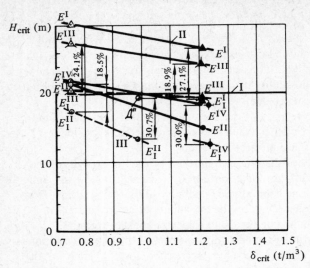

Fig. 46. The relationship $H_{crit} = f(\delta_{init})$. Novy Razdol tertiary clays.
I, average critical height of the waste-heaps in the natural state; II (solid line), before formation of an entire, continuous surface of slip; III (dotted line), after the development of such a surface of slip; E^I and E^{II} are series of models with a rate of deposition $Q \approx 800$ m³/h ($t = 44'48''$) and where $w_{init} \approx 28.2\%$ and $w_{init} \approx 30.9\%$ respectively; E^{III} and E^{IV} are series of models with a rate of deposition $Q = 2400$ ($t = 14'48''$) and where $w_{init} \approx 28.5\%$ and $w_{init} \approx 30.9\%$ respectively.

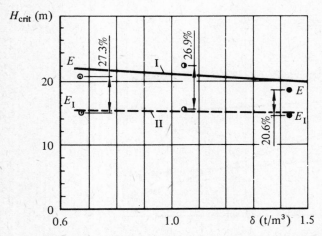

Fig. 47. The relationship $H_{crit} = f(\delta_{init})$. Kerchi tertiary clays. Moisture content $w_{init} \approx 28.4\%$. Rate of deposition of the waste-heap $Q \approx 800$ m³/h ($t = 44'48''$). I, before development of a surface of slip; II, after development of a surface of slip.

Relationship between stability and soil properties

When the initial density of a waste-heap mass was increased by 0.1 t/m³ the critical height of the waste-heaps in our tests were reduced, in metres, by:

For Novy Razdol sandy loams plus clayey loams: by 0.4-1.5.
For Novy Razdol clayey soils: by 1.0-1.5.
For Kerchi clays: by 0.2.

These absolute magnitudes of the reduction in critical height were maintained almost regardless of variations in moisture content (up to 3%) and rate of loading (up to 300%). However, since an increase in moisture content and the rate of loading will by itself reduce the magnitude of the critical height the degree of the influence of the initial density of the soil in the waste-heaps on the critical height will vary in our tests within the limits, in percentage:

For Novy Razdol sandy loams and clayey loams: 2-27.
For Novy Razdol clayey soils: 3-10.
For Kerchi clays: up to 1.

What has been stated is valid in relation to the critical height when defined as earlier. However, in certain cases where the initial angles of the slopes are less than the angle of repose the limit (critical) height of the slope of a compacted or consolidated waste-heap may be greater than the critical height of the slopes of unconsolidated waste-heaps, deposited at the same angles.

As can be seen in the graphs (see figs. 37, 38) the curves for the relationship $\alpha_{lim} = f(H)$ for consolidated waste-heaps are many times steeper than the curves for the same relationship for unconsolidated waste-heaps. In the case of clayey waste-heaps these curves intersect. In all the cases observed these intersections occurred at angles α somewhat (mainly only slightly) less than the angles of repose α_{rep}, as shown in fig. 48.

In models composed of sandy loam and clayey loam no intersections of similar curves were obtained. The curve for $\alpha_{lim} = f(H)$ for consolidated models in these cases reached approximately as far as the ordinate corresponding to the angle of repose $\alpha_{rep} = 35°$, and at this point, close to H_{crit}, it passed over at a steep angle into the curve for $\alpha''_{lim} = f(H)$, i.e. the curve for the relationship as observed after the failure or collapse of the slope (fig. 49).

And since, after the onset of collapse and of fissuring, the physical state of the sliding part of previously consolidated models is scarcely distinguishable in any way from the sliding part of loosely deposited models their curves for $\alpha_{lim} = f(H)$ (after the onset of collapse) practically converge or follow a parallel curve at a short distance apart.

The reduction in the critical height of a waste-heap which occurs with an increase in its initial density can be explained by the more rapid transition of the soil which has been subjected to forced compaction from a three-phase to a two-phase state, and, as a consequence of this, by the earlier development of excess pore pressure which contributes to the failure of that part of the waste-heap contiguous to the slope.

If we compare the holding capacity of the consolidated and loosely deposited

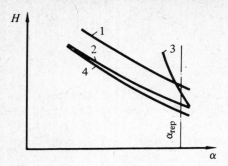

Fig. 48. The character of the curves $\alpha_{lim} = f(H)$ for loose and consolidated clayey waste-heaps.
1, $\alpha_{lim} = f(H)$ with loose deposition of the waste-heap; 2, $\alpha''_{lim} = f(H)$ with loose deposition; 3, $\alpha_{lim} = f(H)$ in a consolidated waste-heap; 4, $\alpha''_{lim} = f(H)$ in a consolidated waste-heap.

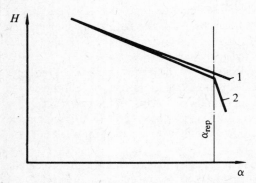

Fig. 49. The character of the curves $\alpha_{lim} = f(H)$ for loose and consolidated sandy loam waste-heaps.
1, with loose deposition of the waste-heap; 2, in a consolidated waste-heap.

waste-heaps at the moment of their failure we shall find that the capacity of the loose waste-heaps exceeds that of the consolidated heaps (even though only slightly).

The influence of the productive rate of waste-heap formation (the rate of deposition)

An increase in the rate of loading of the models by three times, corresponding to an increase in the rate of waste-heap formation in the natural state from 800 to 2400 m³/h, produced a reduction in the critical height of loosely deposited waste-heaps (figs. 50, 51, 52), as a percentage, of:

In the case of Novy Razdol sandy loams plus clayey loams: up to 35, on average by 30.

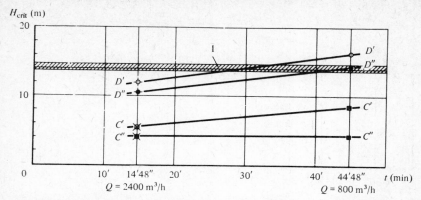

Fig. 50. The relationship $H_{crit} = f(t) = f(Q)$. Novy Razdol sandy loams plus clayey loams. I, average critical values of the height of the waste-heaps in the natural state; C', series of models with loose deposition ($\delta \approx 1.08$ t/m³) with average moisture content $w_{init} \approx 25.0\%$; C'', the same, but with dense deposition ($\delta \approx 1.53$ t/m³) where $w_{init} \approx 25.6\%$; D', the same, but with loose deposition ($\delta = 0.91$ t/m³) and where $w_{init} \approx 22.4\%$; D'', the same, but with dense deposition ($\delta = 1.53$ t/m³) and where $w_{init} \approx 22.4\%$.

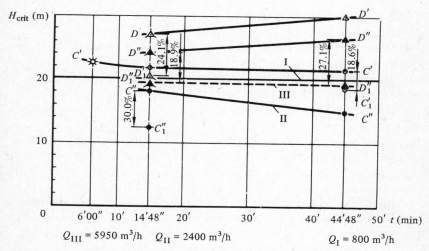

Fig. 51. The relationship $H_{crit} = f(t) = f(Q)$. Novy Razdol tertiary clays. I, average critical height of the waste-heaps in the natural state; II (solid line), before development of a continuous entire surface of slip; III (dotted line), after formation of a continuous surface of slip; C', series of models with loose deposition ($\delta_{init} \approx 0.75$ t/m³) with average moisture content $w_{init} \approx 30.7\%$; C'', the same, but with dense deposition ($\delta_{init} \approx 1.22$ t/m³) and where $w_{init} \approx 30.9\%$; D', the same, with loose deposition ($\delta_{init} \approx 0.75$ t/m³) and where $w_{init} \approx 28.2\%$; D'', the same, with dense deposition ($\delta_{init} \approx 1.21$ t/m³) and where $w_{init} \approx 28.5\%$.

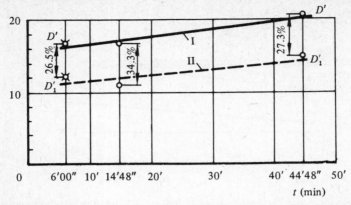

Fig. 52. The relationship $H_{crit} = f(t) = f(Q)$. Kerchi tertiary clays. $Q_{III} = 5950$ m³/h ($t_{III} = 6'00''$); $Q_{II} = 2400$ m³/h ($t_{II} = 14'48''$); $Q_{I} = 800$ m³/h ($t_{I} = 44'48''$); moisture content $w_{init} = 28.4\%$.

I, before development of surface of slip; II, after development of surface of slip; D', series of models with loose deposition ($\delta_{init} = 0.67$ t/m³).

In the case of Novy Razdol clayey soils: up to 10, on average by 5.

In the case of Kerchi clays: by 19.

The extent of the influence of the rate of deposition of a waste-heap in conditions of dense deposition was insufficiently explored. At the same rate of deposition no reduction in the critical height was observed on occasions. For example, in the case of densely consolidated Novy Razdol sandy loams plus clayey loams the average reduction amounted to 15%, but in the case of Novy Razdol clayey soils there was no reduction.

The initial densities of the soils in the waste-heaps observed in the natural state occupy an intermediate position between the 'maximum loose' and 'maximum consolidated' initial states achieved in the models. We are therefore justified in estimating that the middle values of the degrees of influence of the rate of waste-heap formation established for loosely deposited models will be close to the natural state; in other words, given an increase in the rate of deposition from 800 to 2400 m³/h (i.e. by three times) one can expect a reduction in the critical height of the waste-heaps of approximately, in percentage:

For Novy Razdol sandy loams and clayey loams: 20–15.

For Novy Razdol clayey soils: 5–2.

For Kerchi clays: 10.

Two cases of an increase in critical height with an increase in the rate of loading which occurred whilst model testing must be assigned to the category of accidental (possible) deviations.

Accordingly, all three factors which we have considered – moisture content, initial density and rate of deposition of the soils – are of great significance for the determination of the conditions of stability of a waste-heap at the time of its formation and during the initial growth of consolidation. *Our investigations lead to the conclusion that the large scatter of the collated curves representing $\alpha_{lim} = f(H)$, corresponding to the various values of w, δ and Q in figs. 37 and 38 reveals the possibility of serious errors in the determination of the stability of waste-heaps by means of analytical and grapho-analytical calculations which do not take into account the influence of these factors.*

The moisture content of the soil incorporated in a waste-heap cannot be regulated artificially. Regulation of the initial density of the soil in a waste-heap is also improbable given a single chosen system of waste-heap formation. The only means of controlling waste-heaps (to ensure their stability) remaining at the disposal of the technologist are methods of juggling with the productive rate of waste-heap formation, the height of the waste-heap and, more rarely, the possibilities of altering the resulting angles of the deposited waste-heaps. At this point we can regard the results cited of the investigations carried out so far as guide lines for a first approach.

The determination of $H_{crit} = f(w, \delta, Q)$ by the factor experiment method

Employment, for the solution of our problems, of the recommendations of Engineer A. A. Preobrazhensky [16] concerning the use of the statistical method of the factor experiment gave us the opportunity to determine the overall relationship between the stability of unconsolidated waste-heaps and the three simultaneously acting variables: moisture content w, initial density of the soils δ_{init} and their rate of deposition in the waste-heap or, in other words, the productive rate of waste-heap formation Q.

The essence of this method, as distinct from the generally adopted procedure of experimentation, in which the variables under investigation are changed in turn whilst retaining the remaining variables unchanged, lies in the fact that in accordance with a predetermined plan the simultaneous change of all the variables included in the experiment is accounted for. The plan of the experiment is compiled in such a way as to embrace all the possible combinations of the levels of variables (in our case the plan includes eight combinations).

On the basis of the processing of the results of an experiment in accordance with the given plan a so-called mathematical method of the process is compiled, analysis of which enables us to establish the character and the degree of the influence on the magnitude being sought (in our case, on H_{crit}) not only of the separate variables, but also of their combined effect, which is impossible to achieve by the conventional method. In order to exclude uncontrollable external conditions which alter during the process of experimentation (the so-called trends) we adopted the method of randomisation, i.e. individual tests, included in the plan of the experiment, are carried out in random order. As a starting model a linear model of the process was adopted which turned out to be sufficiently convincing and conclusive for our case, as is confirmed by comparison of the calculation data with the natural state.

The plan of the experiment was compiled such that each of the variables under investigation was included in it under two conditions, i.e. maximum and minimum, which correspond to +1 and −1 in the normalised system of variables. In this the plan of the present experiment corresponds to the plan of the factor experiment, type 2^3 (table 26).

The sequence of calculation operations, which remains constant in our cases, is clear from the four examples in appendix 4 of a calculation log completed for the case of the modelling of waste-heaps composed of Novy Razdol tertiary (Tortonian) clays.

The starting equation of the full mathematical model of the process in our examples has the form:

$$H_{crit} = b_0 + b_1 w' + b_2 \delta' + b_3 Q' + b_{12} w' \delta' + b_{13} w' Q'$$
$$+ b_{23} \delta' Q' + b_{123} w' \delta' Q',$$

where $b_0, b_1, b_2, \ldots, b_{123}$ are coefficients of regression, determined by calculation; w', δ', Q' are dimensionless (normalised) values of the variables: moisture content of the soil material, its initial density and rate of deposition in the waste-heap or productive rate of waste-heap formation.

After completion of the calculations and substitution of the dimensionless (normalised) values of the variables by their natural values w, δ, Q the final expressions of the critical height crystallise out by means of the exclusion of the statistically unreliable terms into calculation formulae which are convenient for practical use and which define not only the direct relationships between the critical heights of the waste-heaps and the three main factors (variables) but also their interaction with each other.

For the two types of Novy Razdol waste-heaps investigated the final calculation formulae for the relationship $H_{crit} = f(w, \delta, Q)$ take the following form:

For waste-heaps of quaternary soils (sandy loams plus clayey loams) in metres:

$$H_{crit} = 103.22 - 3.41w - 7.31\delta - 0.01663Q + 0.00197\delta Q$$
$$+ 0.00052wQ;$$

for waste-heaps of tertiary (Tortonian) clays, in metres:

$$H_{crit} = 177.27 - 4.83w - 9.08\delta - 0.031Q + 0.001wQ,$$

where w is the moisture content by weight of the soils, in per cent; δ is the initial density of the soils in the waste-heap, expressed in terms of the bulk weight of the skeleton, in t/m^3; Q is the productive rate of waste-heap formation, in m^3/h.

The verifications of the concurrence of the 'actual' (found by modelling) values $H_{crit\ (act)}$ and the calculated values $H_{crit\ (calc)}$, obtained using the derived formulae, are set out in tables 27 and 28.

Table 26 *Plan of the factor experiment in a normalised system of variables*

Test number	Model numbers	Moisture content w (%)		Density δ (g/cm³)		Rate of waste-heap formation Q (m³/h)		Critical height of waste-heap H_{crit} (metres)	Calculation index
		Actual value	Normalised value	Actual value	Normalised value	Actual value	Normalised value		
1	N'_1–N'''_1	w_1	−1	δ_1	−1	Q_1	−1	$H_{crit\,1}$	y_1
2	N'_2–N'''_2	w_2	+1	δ_2	−1	Q_2	−1	$H_{crit\,2}$	y_2
3	N'_3–N'''_3	w_3	−1	δ_3	+1	Q_3	−1	$H_{crit\,3}$	y_3
4	N'_4–N'''_4	w_4	+1	δ_4	+1	Q_4	−1	$H_{crit\,4}$	y_4
5	N'_5–N'''_5	w_5	−1	δ_5	−1	Q_5	+1	$H_{crit\,5}$	y_5
6	N'_6–N'''_6	w_6	+1	δ_6	−1	Q_6	+1	$H_{crit\,6}$	y_6
7	N'_7–N'''_7	w_7	−1	δ_7	+1	Q_7	+1	$H_{crit\,7}$	y_7
8	N'_8–N'''_8	w_8	+1	δ_8	+1	Q_8	+1	$H_{crit\,8}$	y_8

Table 27 Verification of concurrence of actual and calculated values of H_{crit} (Novy Razdol sandy loams plus clayey loams)

Test number	Model numbers	H_{crit} (act) (m)	H_{crit} (calc) (m)	ΔH_{crit} = H_{crit}(act) − H_{crit}(calc) (m)	Relative calculation error (%)	$(\Delta H_{crit})^2$	Anticipated error in calculation	
							Mean square	Maximum
1	24–25	16.3	17.50	−1.20	7.3	1.44	$S_{H_{crit}} = \sqrt{\left(\dfrac{\Sigma(H_{crit}(act) - H_{crit}(calc))^2}{n-1}\right)}$	$\Delta_{max} = 2S_{H_{crit}}$
2	22–23	8.7	9.05	−0.35	4.0	0.12		
3	26–27	14.3	13.33	+0.97	6.7	0.94		
4	64–65	4.5	4.27	+0.23	5.1	0.05	$\sqrt{\left(\dfrac{3.82}{8-1}\right)} \approx 0.74$ m	2. 0.74 = 1.48 m
5	56–57	12.0	12.28	−0.28	2.3	0.08		
6	58–59	5.6	5.98	−0.38	6.7	0.15		
7	60–61	10.5	10.65	−0.15	1.4	0.02	or ~7.8%	or ~15%
8	62–63	4.3	3.29	+1.01	23.4[a]	1.02		
				−2.36 +2.21	Average ~7.1	Σ = 3.82		

[a] The large relative error in test 8 is explained by the significant deviation in the moisture content of models 62–63 from the norm established in the experiment and by the small height of the waste-heap.

Table 28 Verification of the concurrence of the actual and calculated values of H_{crit} (Novy Razdol tertiary clays)

Test number	Model numbers	H_{crit} (act) (m)	H_{crit} (calc) (m)	ΔH_{crit} = H_{crit} (act) − H_{crit} (calc) (m)	Relative calculation error (%)	$(\Delta H_{crit})^2$	Anticipated error in calculation	
							Mean square	Maximum
							$S_{H_{crit}} = \sqrt{\dfrac{\Sigma(H_{crit}\,(act) - H_{crit}\,(calc))^2}{n-1}}$	$\Delta_{max} = 2 S_{H_{crit}}$
1	34–35	30.1	32.0	−1.9	6.3	3.61		
2	36–37	21.5	21.5	0	0	0		
3	54–55	26.5	27.4	−0.9	3.4	0.81	$\sqrt{\left(\dfrac{15.05}{8-1}\right)} \approx 1.47$ m	$2 \times 1.47 = 2.93$ m
4	40–41	15.0	16.7	−1.7	11.3	2.89		
5	52–53	27.3	27.3	0	0	0		
6	44–45	21.8	20.7	+1.1	5.0	1.21	or ~6%	or 12.7%
7	50–51	24.3	22.1	+2.2	9.1	4.84		
8	48–49	18.2	16.9	+1.3	7.1	1.69		
				−4.5 +4.6	Average 5.3	Σ = 15.05		

The factor experiment method enables us to find the minimum critical height of a waste-heap with guaranteed accuracy. The maximum anticipated error Δ_{max} (where $P = 0.95$) is equal to twice the mean square error

$$\Delta_{max} = 2 \sqrt{\left(\frac{\Sigma(H_{crit\,(act)} - H_{crit\,(calc)})^2}{n-1}\right)},$$

where n is the number of tests (in the present case $n = 8$). The guaranteed value of the critical height of the waste-heap is determined by calculation in accordance with the empirically obtained relationship $H_{crit} = f(w, \delta, Q)$ reduced by the magnitude of the anticipated maximum error Δ_{max}.

4

The influence of the composition of a soil mixture on the stability of an unconsolidated waste-heap

The soil mixture of the waste-heap

Amongst mine-managers and scientific workers the opinion is commonly held that the stability of hazardous waste-heaps of clayey soils can be increased by increasing the proposition of the sandy component in the waste-heap soil mixture. At the same time, however, it is known from practical experience in open workings, and confirmed by the tests described earlier, that in the majority of cases, at the magnitudes of moisture content observed in natural conditions, pure sands are more stable in waste-heaps than clays, and clays are more stable than poor clayey loams and sandy loams. The latter may be regarded, with a certain tolerance, as well mixed fine-grained sands with clays.

When comparing the results of the determination of the critical heights of waste-heaps H_{crit}, described earlier, for soils of different lithologic composition a paradoxical phenomenon is revealed:

H_{crit} for sand $> H_{crit}$ for clays $> H_{crit}$ for sand plus clays.

To find an answer to the question of the reason for, and the boundaries of the influence of this phenomenon model waste-heaps were tested with varied composition of the basic mixed soils: Kerchi tertiary soils, heavy clays and Dnieper sands from the Kiev shore, the physico-mechanical properties of which are set out in table 6 and in figs. 18 and 19.

Before starting the experiments it was necessary to make a prior choice between models prepared from soils taken directly from actual waste-heaps and models constructed from carefully mixed soils of variable composition, taken from mining sites in predetermined proportions of the components: clay and sand.

The soil composition of an actual mixed waste-heap changes continuously in any direction of a section of the waste-heap. The elements of chance in the composition of the soil mixture are so obvious in this case that they do not permit the solution of a general problem on the basis of a particular example of an actual waste-heap. In addition, the possibilities of constructing an authentic model of chaotically deposited mixed waste-heaps are problematical. At the same time the laws governing the influence of the composition of soil mixtures on the stability of a waste-heap, which are more easily established in artificial mixtures, are applicable within wide limits to actual waste-heaps. It was therefore to this direction that we gave preference when arranging tests with model waste-heaps with variable composition of mixed soils.

The determinations of the shear resistance and stability of mixtures of clays with sands in waste-heaps already have their own history. In the laboratory at Ukr.N.I.I.Projekt concerned with the stability of (earthen) faces and the drainage of open-cast mines investigations of this kind have been carried out twice: in 1961 and in 1967. Their results were published in the collected papers [2 and 19]. In the first case mixtures were made of lower Neogene clays and yellow-brown clayey loam from the Balakhov lignite mine. As a result of the tests they came to the following conclusions.

Soil mixtures of the first group, containing lower Neogene clays in amounts of 22% and above, are indistinguishable from pure clays in respect of shear resistance. Soil mixtures of the second group, containing 5% and above of sandy Neogene clays are indistinguishable in respect of shear resistance from pure sandy clays. In the course of these experiments it remained unnoticed that with the growth of the compressive loadings the influence of the active constituent – clay – on the strength of the mixture increases appreciably.

In the second case the mixtures were constituted of variable proportions of percarbonic sands and clays from the Svobodnensk coal deposits. The determinations were carried out with tests repeated from three to five times. The results of the determinations are set out in fig. 53. These investigations may be summarized as follows:

In waste-heaps containing up to 10% of clays the stability of the waste-heap masses is determined by the strength of the sands. In clayey waste-heaps containing up to 30% of sand the stability of the waste-heaps is determined by the strength of the clays. The remaining mixtures occupy a position intermediate between those for sandy and clayey waste-heaps.

In both the cases described the moisture contents of the soil components remained constant – equal to their average magnitudes in a natural deposit (and were therefore random moisture contents).

Interesting data concerning the strength of mixtures (lumps) of clayey soils combined with fine-grained sands were published in 1967 by H. Schneckenberg [22] (G.D.R.) in connection with investigations into the reasons for the reduction in the stability of waste-heaps when the waste-heap soils were transported by conveyer. H. Schneckenberg determined the strength of six soil mixtures with different contents of fine-grained sand in two series of tests. Series B corresponded to determinations of the strength of the mixtures after they had been processed in advance on a special vibration machine which imitated the shaking of the soil on a conveyer belt: series O was for the same mixtures, but without

Influence of composition of soil on stability

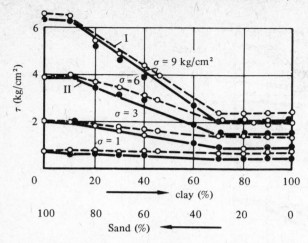

Fig. 53. Graphs of the relationship between the shear resistance of soil mixtures (clay plus sand) and their compositions (according to T. D. Ustinova).
I, average values at natural moisture content of the soil mixtures; II, average values with capillaric wetting of the mixtures.

passing them through the vibratory equipment.

As in the case of the tests already described by Ukr.N.I.I. Projekt, the 'natural' moisture content of the soils was accepted, i.e. accidental, but constant for each component soil in the mixture. The results of the tests are given in fig. 54.

The mixtures subjected to vibration showed a sharp reduction in strength when their sand content was increased to 40%, whilst the samples not subjected to vibration (of the same composition) did not reveal this and their strength gradually increased with an increase in the sand content.

H. Schneckenberg considers that the reasons for this phenomenon are a change in the consistency of the cohesive components in the soil mixture and a deterioration in the filtration properties on account of the process of the mixing of the components at the transfer points on the conveyor. These reasons evoke no objections, but are inadequate since they themselves demand an explanation of the reasons for the change in consistency and for the deterioration in the filtration properties of the mixture. We find the answer in the very arrangement of the test. In essence two different mixtures are being compared: in the one case the spaces between the large lumps of the cohesive clayey soil are filled by moist sand, whilst in the second case the crushed fine aggregates of the cohesive soil enter into more intimate contact with those same moist sands. In the first case there is no means of access for the water from the sands into the interior zones of the clayey lumps, but in the second case free migration of the water from the sands into the clays occurs.

In our tests we are pursuing the aim of establishing the maximum extents of the influence of concrete factors on the stability of a waste-heap slope; in

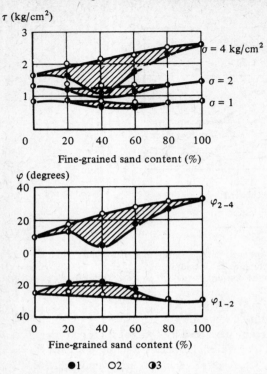

Fig. 54. The influence of the percentage content of fine-grained sand and of its moisture content on the shear resistance of a soil mixture when transported by conveyer (according to H. Schneckenberg).
Moisture content: mixture 26%; sand 28–30%; clay 23.6–24.9%. τ (kg/cm^2) = shear resistance; φ (degrees) = angle of internal friction; 1, tests of series B; 2, tests of series O; 3, results coinciding in both series.

particular, the effect of an inert sandy addition to clayey soils in waste-heap mixtures.

We shall consider this problem in two variants: when the possibility of migration of water from the sands into the clays is excluded and when such migration is allowed to occur.

To permit comparison between the results obtained from the new tests with the preceding investigations of the influence of the moisture content of waste-heap soils, the moisture content of the clayey material (Kerchi tertiary clays) and the calculated productive rate of waste-heap formation are kept the same as in the basic test models, i.e. $w_{clays} = 28.4\%$ and $Q = 800$ m^3/h.

The preceding tests established the limit values of the moisture content of Dnieper sands, corresponding to a maximum molecular moisture capacity $w_{mol} = 3.05\%$ and an average capillaric moisture content under a compressive load of $\sigma = 3$ kg/cm^2 equal to $w_{cap} = 13.6\%$.

Influence of composition of soil on stability

In the first case the pellicular water in the sand is held so firmly that it is not torn away from the sand by centrifuging even when accelerations of the order of $300\,g$ are achieved, where g is the acceleration of the force of gravity.

When sands with this moisture content are added to clayey soils migration of water from the sands into the clays is excluded, and the results of comparative tests of similar mixtures of clays with sands reflect only the effect of an inert addition (sand) on the stability of a waste-heap, the moisture content of whose clayey and sandy components is maintained constant.

In the second case the water contained in capillarically wetted sands can easily pass into the much finer capillaries of the clay in an amount w_{free}, equal to the difference between the values of the capillaric and maximum molecular moisture capacities of the sands. In our case

$$w_{free} = w_{cap} - w_{mol} = 13.60 - 3.05 = 10.55\%.$$

In a mixture of sands plus clays ($S + C$, where S is the weight of the skeleton of the sand, in kg, and C is the weight of the skeleton of the clay, in kg) the free water which can pass from the sands into the clays is defined as, in kg:

$$\frac{w_{free} \times S}{100}$$

and the additional wetting of the clays, in percentage, as:

$$\frac{w_{free} \times S}{C}.$$

This being so, the maximum possible moisture content of the clayey component in the mixture $w_{C\,max}$ amounts to:

$$w_{C\,max} = w_C + \frac{w_{free} \times S}{C},$$

where w_C is the initial moisture content of the clays, in our case equal to 28.4%.

In order to obtain comparable magnitudes and to exclude accidental influences the soil components were mixed together using the identical technological procedure. The sands were mixed with the clays in an air-dry condition. The clays were previously ground in a disc mill to an aggregate size of not more than 0.01 mm. After uniform mixing of the components the mixture thus obtained was wetted with the aid of a spraying machine and its moisture content brought to an average calculated level, in percentage, of:

$$w_{av} = \frac{S \times w_S + C \times w_C}{S + C}.$$

In doing this, allowance was made for the initial moisture contents of the air-dry components. The mixtures prepared in this way were then passed through a rubbing machine and packed in hygrostatic troughs, where they were kept for 24 hours before being incorporated in the models.

The moisture content was controlled both in the hygrostatic troughs and directly during the deposition of the soil mixtures in the models. In both cases test samples of no less than two batches from every 10 kg of the soil mixture were taken.

The models were constructed by the method of loose deposition 'from the shovel'.

Because of the inconstancy of the bulk weight of the mixtures the magnitude of the initial bulk weight of the skeleton was not fixed in advance, but merely recorded according to the actual state of the model before starting to load it (before acceleration).

The plan of the experiment

The plan of the experiment to determine the degree of influence of sandy additions on the critical height of a clayey waste-heap, in conditions of molecular and capillaric wetting of the sands, was constructed in accordance with the most simple scheme of changing the percentage content of the sandy component S in the overall weight $C + S$ of the skeletal material of the waste-heap soil mixture.

$$H_{crit} = f\left(\frac{S}{C+S} \times 100\right).$$

As a rule, the interval selected between the varying contents of the sandy component of the mixture was 10%, but at certain stages this was altered with the object of making the character of the linearity of the functions more precise.

When arranging the experiment, and with the purpose of reducing the laborious work of converting and classifying the components of the sand and clay according to their contents of the relevant fractions of the elementary grains, it was decided to allow the components of the mixtures to be determined in a more simplified way, namely: S represents sands, as the overall complex of the fractions comprising the Dnieper sands from the Kiev shore (see table 6) and C represents clays, as the complex of fractions comprising Kerchi tertiary heavy clays (see table 6).

The plan of the experiment embraced eighteen models, including nine for cases of molecular wetting of the sands (where $w_S = 3.05\%$) and nine for cases of capillaric wetting of the sands (where $w_S = 13.60\%$).

The initial moisture content of the clayey component was maintained constant, and equal to $w_C = 28.4\%$, in all tests.

A composite plan of the experiment, accompanied by data of certain calculation and actual magnitudes which characterized the models included in the plan, is given in table 29.

The character of the variations in shear resistance of soil mixtures in relation to their composition

All the soil mixtures used in the construction of the models were subjected to shear resistance tests.

Each mixture was tested twice: at the initial moisture content of the soil mixture at the moment of its incorporation in the model and in conditions of free capillaric wetting (to total saturation or expulsion) after the application to the sample of a compressive load equal to that under which shearing occurs.

Since almost every mixture served twice as the material for the models, in conditions of molecular and capillaric wetting of the sands, in the various conditions of wetting each soil mixture was subjected to four shear tests, with the compilation of four data-sheets for strength.

It would seem that identical soil mixtures, afforded the opportunity for free capillaric wetting under identical compressive loads, should possess identical strength regardless of the extent of their original wetting; and this, evidently, should be valid also in the case of a lengthy dwell in identical conditions of loading and opportunity for wetting. In practice, however, the process of saturation of soil with water under load, which in the beginning proceeds quite intensively, subsides after several days and beyond that continues so slowly, that in the real time conditions of the advance of an open waste-heap it can be regarded as having ceased.

In the conditions of laboratory tests the process of swelling by a sample is normally considered as ended when the increase in the height of the specimen h during a 24-hour period does not exceed $0.0003\ h$.

As a result of this the capillarically wetted samples of identical mixtures, which had different initial moisture contents, during the periods of time of interest to us, differ from each other, even though only slightly, in moisture content and strength.

The derived relationships of the shear resistance of mixtures of Kerchi clays and Dnieper sands are shown, after processing, in the graphs set out in figs 55 and 56 within the coordinates of the shear resistance τ, in kg/cm^2, and the percentage content of the sandy component of the mixture ($100S/(C + S)\%$).

The results obtained differ substantially from the data cited earlier by earlier investigators of the strength of soil (waste-heap) mixtures.

Let us consider three possible conditions of the wetting of a waste-heap mass.

First Case. Waste-heaps of clay with a certain constant moisture content (in our case $w_C = 28.4\%$) are made leaner by the addition of fine-grained river sand, whose moisture content does not exceed the molecular maximum (3.05%). In this case the additions of sand in small quantity produce no substantial effect on the strength of the waste-heap mass; above a certain limit, however, dependent on the compressive load acting on the soil mass, the effect of the

Table 29 *Composite plan of the experiment with model waste-heaps of variable com
clays and of the capillaric water introduced by the sands*

Moisture content (%)	Composition of skeleton (kg) $\frac{clay}{sand}$			$\frac{100}{0}$	$\frac{90}{10}$
Clays w_C = 28.4 Sand w_S = w_{mol} = 3.05	Model numbers			66–67	120
	δ_{init} (g/cm³)			0.67	0.7:
	w_{av} (%)	Calculated		28.4	25.84
		Actual		28.4	25.9
	H_{crit} (metres) as determined by test			20.5	31.2
Clays w_C = 28.4 Sand w_S = w_{cap} = 13.6 Free moisture content of sand w_{free} = w_{cap} − w_{mol} = 10.55	Model numbers			66–67	107–
	δ_{init} (g/cm³)			0.67	0.7!
	w_{av} (%)	Calculated		28.40	26.9!
		Actual		28.4	26.6–
	Calculated indices	Free water $w_{free} \times S/100$ (kg)		0	1.06
		Additional moisture content of the clays $w_{free} \times S/C$ %		0	1.1!
		Maximum possible moisture content of the clays $w_C + (w_{free} \times S/C)$ %		28.40	29.5!
	H_{crit} (metres) as determined by test			20.5	24.9

[a] Transition to a stepped form of collapse.
[b] Protrusion in the vicinity of the surface of the slope.

addition of sand becomes noticeable and manifests itself in the steadily increasing strength of the mixture. With an amount of sand close to 90% (on the overall weight of the skeleton of the mixed soils) the strength of the mixture reaches a maximum, close to and even exceeding the shear resistance of pure sand.

The limit values of the minimum amount of sand at which the strength of the mixture remains almost identical to the strength of pure clays will vary within wide limits. The greater the compressive load, the lower is this limit. Thus in our case, at compressive loadings $\sigma \leqslant 1$ kg/cm², this limit is scarcely perceived at a sand content of about 90%. Where σ = 3 kg/cm² it equals 50%, but where σ = 6 kg/cm² it reaches 35%. Evidently, in the given case the development of pore pressure in the clayey component of the mixture is making itself felt, as also the effect of the sandy addition on its reduction.

Influence of composition of soil on stability

ition to determine the effect on the critical height of a waste-heap of a sandy addition to

$\frac{70}{30}$	$\frac{65}{35}$	$\frac{60}{40}$	$\frac{50}{50}$	$\frac{30}{70}$	$\frac{20}{80}$	$\frac{10}{90}$	$\frac{0}{100}$
104–112	121	113	102	100	–	93	82–83
0.81–0.83	0.78	0.81	1.00	0.81	–	1.11	1.41
20.81	19.52	18.24	15.73	10.65	–	5.58	From 13.6 in the beginning to 3.05 at the end
21.2	19.7	18.1	15.5	10.60	–	5.1	
34.0[a]	29.7	33.0	(Did not fail) > 50.5	(Did not fail) > 51.3	–	(Did not fail) > 55.1	(Did not fail) > 51.8
105	122	–	111	101	110	99	82–83
0.79	0.82	–	0.99	1.35	1.37	1.54	1.41
23.97	23.21	–	21.00	18.04	16.57	15.07	From 13.6 in the beginning to 3.05 at the end
23.4	23.5	–	20.7	18.1	16.9	14.4	
3.17	3.69	–	5.28	7.39	8.45	9.50	From 10.55 at start to 0 at end
4.53	5.68	–	10.55	24.60	42.20	95.00	–
32.93	34.08	–	38.95	53.00	70.60	123.40	–
24.1	12.5	–	11.2	4.2	34.4[b]	(Did not fail) > 61.5	(Did not fail) > 51.8

Second Case. Sands used to make a clayey waste-heap more lean have been wetted capillarically (w_S = 13.6%). This corresponds to the majority of cases involving open-cast coal mines of the central and northern belts. The extreme parts of the curve of the relationship $\tau = f\,[100S/(C + S)]$ are in this case similar to those in the preceding one. But at the boundary between the second and third thirds of the curve a sharp reduction in the strength of the mixture is observed on account of the additional wetting of the clays by capillaric water passing from the sands. The position of extremity is reached in our case at approximately 70% sand component.

H. Schneckenberg observed a similar extremity at 40% of sands with vibration of clayey waste-heap mixtures. However, in the absence of vibration no extreme point was reached in his tests. Evidently, this must be explained by inadequate contact between the large lumps of clayey soil and the sands.

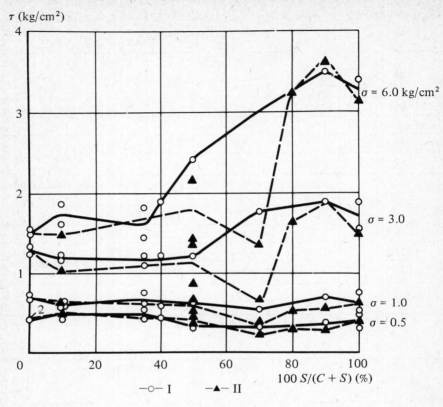

Fig. 55. The relationship between the shear resistance τ of mixtures of clays C with sands S and the percentage content of the sandy component ($100\,S/(C+S)$) with no opportunity for additional wetting.

I, initial moisture content of the sand 3.05%; II, the same, at 13.6%. In both cases the initial moisture content of the clay was 28.4%.

Third Case. Free wetting of the waste-heap mixtures, a lengthy period of time resting in a loaded condition under their own weight and open to the ingress of underground waters.

Additional capillaric wetting of the mixtures, composed of clays with moisture content 28.4% and previously capillarically wetted sands (up to $w_S = 13.60\%$) is hardly reflected at all in the strength of the mixtures, with the exception of small compressive loads (up to $\sigma = 3.0$ kg/cm^2), when the clayey component has the opportunity for certain additional swelling.

Mixtures involving practically dry sands, having an initial moisture content up to 3.05% and afforded the opportunity for additional capillaric wetting, maintain a tendency towards a reduction in strength even at higher compressive loadings, although within only slight limits.

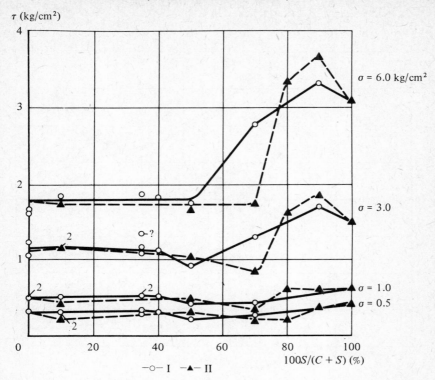

Fig. 56. The relationship between the shear resistance of mixtures of clays C with sands S and the percentage content of the sandy component ($100 S/(C + S)$) with additional capillaric wetting of the mixtures.
I, initial moisture content of the sand 3.05%, II, the same, at 13.6%. In both cases the initial moisture content of the clay was 28.4%.

The relationship between the critical heights of waste-heaps, the critical angles of the slopes and character of the failure of the waste-heaps on the one hand, and their soil composition on the other

The relationships between the critical angles of the slopes and the heights of waste-heaps, in the form of integrated graphs of $\alpha_{lim} = f(H)$ for different compositions of mixtures of Kerchi tertiary clays and Dnieper sands, are set out in fig. 57.

In one series of tests sands were used in the mixtures which had a moisture content corresponding to their maximum molecular moisture capacity, i.e. where $w_S = 3.05\%$. In another series the moisture content of the sands included in the mixtures was equal to their average capillaric moisture content, i.e. $w_S = 13.60\%$. The starting moisture content of the clays remained in both cases at a constant: $w_C = 28.4\%$.

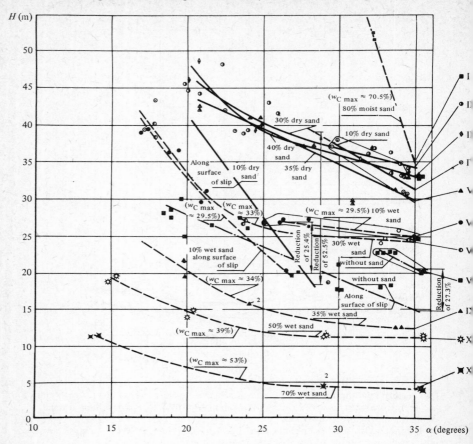

Fig. 57. The relationship between the critical angles of the slopes α_{lim} and the height of mixed waste-heaps H. Mixtures of Kerchi tertiary clays with Dnieper sands.

I, model 110 (protruding) (H_{crit} = 34.4 m; $C:S$ = 20:80; w_{init} = 16.9%; δ_{init} = 1.365 t/m³); II, models 104–112 (transition to jerky, ridged form of failure) (H_{crit} = 34.0 m; $C:S$ = 70:30; w_{init} = 21.0–21.2%; δ_{init} = 0.814–0.831 t/m³); III, model 113 (H_{crit} = 33.0 m; $C:S$ = 60:40; w_{init} = 18.10%; δ_{init} = 0.810 t/m³); IV, model 120 (H_{crit} = 31.2 m; $C:S$ = 90:10; w_{init} = 25.9%; δ_{init} = 0.724 t/m³); V, model 121 (H_{crit} = 29.7 m; $C:S$ = 65:35; w_{init} = 19.7%; δ_{init} = 0.78 t/m³); VI, models 107–123 (H_{crit} = 24.8 m; $C:S$ = 90:10; w_{init} = 26.6–26.8%; δ_{init} = 0.706–0.707 t/m³); VII, models 105 (H_{crit} = 24.1 m; $C:S$ = 70:30; w_{init} = 23.4%; δ_{init} = 0.793 t/m³); VIII models 66–67 (H_{crit} = 20.5 m; $C:S$ = 100:0; w_{init} = 28.4%; δ_{init} = 0.672 t/m³); IX, model 122 (H_{crit} = 12.5 m; $C:S$ = 65:35; w_{init} = 23.5%; δ_{init} = 0.815 t/m³); X, model 111 (H_{crit} = 11.2 m) $C:S$ = 50:50; w_{init} = 20.7%; δ_{init} = 0.985 t/m³); XI, model 101 (H_{crit} = 4.2 m; $C:S$ = 30:70; w_{init} = 18.1%; δ_{init} = 1.348 t/m³).

Influence of composition of soil on stability

Fig. 58 presents in graphic form the relationship between the stability of waste-heaps, expressed in terms of the critical height of the waste-heap, and the content of the sandy component in them, $H_{crit} = f[100 S(C + S)]$.

The two branches of the curves correspond to the two extreme limits of the possible initial wetting of the sandy component of the mixture ($w_S = 3.05\%$ and $w_S = 13.6\%$). The lower curve corresponds to the case of the free swelling of the clay component on account of the passage of water into the clays from the capillarically wetted sands.

In the practical reality of waste-heap formation the extreme cases do not always occur. In the majority of cases one should expect the critical height of the waste-heaps to be in the interval between the upper and lower limits.

The curves obtained in fig. 58 show poor agreement with the relationship between the strength and the composition of the soil mixture (see figs. 55 and 56). Nevertheless, the points of the stability curves found experimentally lie in a regular order, revealing the established relationship between the stability of a soil mixture and its composition, and the extreme points of the turn in the curve at the minimum level of stability, with a 70% content of capillarically wetted sand in the mixture, coincide with the experimental determinations of the minimum strength of the mixtures.

For mixtures containing capillarically wetted sands the reduction in the

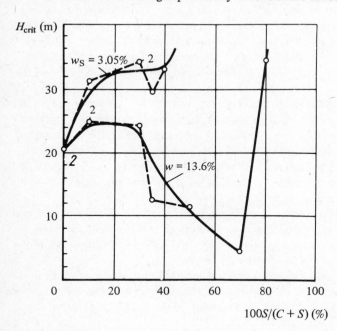

Fig. 58. The relationship between the critical height of a clayey waste-heap H_{crit} and the percentage content of a sandy addition ($100 S/(C + S)$).

stability of the mixed waste-heaps with 70% sandy component reaches 80% by comparison with a waste-heap of pure clay. The critical height of the waste-heap is reduced from 20.5 to 4 metres. The certain increase in the stability of a waste-heap with a content of 10–30% of capillarically wetted sandy component, which was revealed in our tests, may be accidental and therefore requires confirmation in a larger number of tests.

For mixtures of clays with sands whose moisture content lies within the limits of the maximum molecular moisture capacity, a sandy addition increases the critical height of a waste-heap at all stages.

The lack of agreement between the values for stability obtained in the models and the strengths of the corresponding materials of the models as determined in the laboratory requires confirmation in a greater number of tests than have been carried out so far.

If the strength of soil mixtures and the critical height of waste-heaps composed from them is determined over a comparatively wide range of mixture compositions by the strength of the clayey component, the character of the failure or collapse of a model waste-heap depends to a large extent on the sand content in the mixture.

The presence of well-defined continuous and complete surfaces of slip is characteristic of failures of waste-heaps of clayey soils. With the development of continuous surfaces of weakening there occurs a sharp reduction in the stability of a waste-heap, which is no longer capable of maintaining the original critical height of the slope or embankment. This characteristic of the failure of a slope changes with the addition to the clays of a waste-heap of a comparatively small amount of sand (of the order of 30% of the overall weight of the skeleton of the mixture).

With a further increase in the sand content the failure of the waste-heap takes on a 'ridged, jerky' character*. It proceeds in steps, superimposing themselves one on another in the form of weakly advancing faults, which arise as a result of the collapse of potentially accumulated loads from the weight of the deposited soil masses, which are maintained between collapses primarily by frictional forces. With each new collapsing (protruding) step the resulting angle of the slope gradually advances.

From the point of transition to a 'ridged, jerky' character of failure the curve for the relationship $\alpha_{lim} = f(H)$ maintains a uniform level, i.e. with repeated shifts and slips of the waste-heap no reduction in the limit (critical) angles of the slope at the same height of the waste-heap is observed.

The single characteristic protrusion (or bulging) of a slope was observed in model 110 composed of the mixture $C/S = 20/80$ with an initial moisture content of the sand $w_S = 13.60\%$. The critical height up to the time of the start of the bulging-out of the slope reached 34.4 m.

*This term was introduced by us by analogy with the character of the structure of the Skiborge Mountains in the Carpathians.

Influence of composition of soil on stability

At a lower moisture content of the sands and greater content of the sandy component we did not succeed in producing failure of the models even under loads equivalent to a waste-heap height of the order of 50-60 metres.

Conclusions from the tests with soil mixtures

Despite the inconsistencies, which as yet remain unexplained, in the data obtained experimentally (in a first approach) concerning the strength and stability of soil mixtures in waste-heaps, the following practical conclusions can be drawn on the basis of the tests carried out:

1. It should be considered as conclusively demonstrated that additions to clayey waste-heaps of capillarically wetted or capillarically potentially wettable fine-grained sands in certain proportions by weight to the overall weight of the skeleton of the whole mixture (in our tests, from 30 to 80%) can sharply reduce the stability of waste-heaps. In our cases the lowest stability was possessed by mixed waste-heaps which included capillarically wetted sands in amounts of 60-70% of the weight of the overall skeletal mass of the mixture. The stability of these waste-heaps was 5-8 times lower than that of waste-heaps of pure clays.

2. Additions of air-dry pure sands at a moisture content of the order 3-4% increase the stability of clayey waste-heaps.

3. The stability of waste-heaps composed of clays well mixed with sands, with a sand content of less than 30% by weight on the overall weight of the skeleton of the mixture, is determined predominantly by the properties and stability of the clays and, conversely, where the sand content in the mixture is in an amount greater than 80%, the stability of a mixed waste-heap is determined by the stability of the sand.

4. Waste-heaps formed by the simultaneous deposition of clays and fine-grained sands without uniform mixing may possess different strengths in different parts of the body and their stability, depending on the local composition of the components and their moisture content, can vary from high to low.

5. In the examples of the tests carried out with models composed of pure Dnieper sands, Novy Razdol sandy loams plus clayey loams, Novy Razdol Tortonian and Kerchi tertiary clays it can be seen that their strength and stability in waste-heaps is in the same degree as in mixed waste-heaps, and depends on their lithologic and, of primary importance, their granulometric compositions and their wetting.

6. The migration of water in a waste-heap depends on the filtration properties of the waste-heap mass. In models 82 and 83 of a waste-heap composed of pure Dnieper sands with an initial moisture content $w_s = 13.60\%$, the moisture content had dropped by the end of the test to 3.05% on average, i.e. to the limits of the maximum molecular moisture capacity.

With the addition of a clayey component, given thorough mixing of the mixtures, the coefficient of filtration is sharply reduced and the migration of water in the waste-heap is also rapidly decelerated. In table 30 comparative data are given of the initial and final moisture content at six predetermined levels in the models, constructed from mixtures of different composition. From the table it follows that there is practically no migration of water in the models when the initial moisture content of the sand is close to the maximum molecular moisture capacity.

The expulsion of capillaric waters from models constructed from mixed soils occurs in a very limited amount and with a clay content of only 10% in the mixture the difference between the average moisture content of the samples from the upper and lower levels amounts to less than 0.5%.

Table 30. The final distribution of moisture content w, in percentage, in models of mixed waste-heaps (Kerchi clays plus Dnieper sands) after tests in the centrifuge

Diagram showing disposition of test samples over the height of the models.

Composition of the soil clay/sand	100/0	90/10	70/30	70/30	65/35	50/50	30/70	20/80	10/90	0/100
Model numbers	114	106	107	104	105	108	109	102b	111	100	101	110	99	83	82
w_{init} (%)	28.0		26.6	21.2	23.4			15.5	20.7	10.6	18.1	16.9	14.4	12.8	13.8
Test sample numbers															
Test sample 1A						19.3	24.3	15.1	21.0	11.0	17.3	15.3	10.2	2.4	2.8
1B	27.5					19.3	24.4	16.8	20.4	10.7	17.7	15.4	10.7	2.9	2.9
Average w_{final}						19.3	24.4	16.0	20.7	10.8	17.5	15.4	10.4	2.6	2.8
Test sample 2A		24.9	26.4	20.1	23.3	20.1	25.0	15.4	21.2	10.5		13.6	6.0	11.6	
2B		24.8	26.4	20.9	23.5	19.8	24.9	16.7	20.9	10.5		14.2	6.0	11.5	
Average w_{final}		24.8	26.4	20.5	23.4	19.3	25.0	16.0	21.0	10.5		13.9	6.0	11.6	
Test sample 3A		26.8	26.4	21.0	23.7	20.1	25.0	15.3	19.3	10.1	17.5	14.3	5.8	11.6	3.0	2.9
3B		25.2	26.5	21.0	22.6	19.8	24.9	15.5	20.0	10.1	17.8	13.7	5.4	11.9	3.2	3.5
Average w_{final}		26.0	26.4	21.0	23.2	20.0	25.0	15.4	19.7	10.1	17.6	14.0	5.6	11.8	3.1	3.2
Test sample 4A		24.9	26.7	20.8	22.3	19.6	24.5	15.4	20.4	10.6	17.4	13.5	5.0	11.6	3.2	4.1
4B	27.5	25.0	26.8	20.6	22.7	19.7	24.4	15.4	20.5	10.4	17.3	14.1	5.0	11.8	3.3	3.3
Average w_{final}		25.0	26.8	20.7	22.5	19.6	24.4	15.4	20.4	10.5	17.4	13.8	5.0	11.7	3.2	3.7
Test sample 5A		24.8	26.6	20.5	23.9	20.2	25.9	16.3	20.7	11.0	17.2	13.3	5.2	11.8	3.4	3.0
5B	27.8	24.9	26.8	20.7	23.0	19.9	26.2	15.9	20.5	10.5	16.8	13.0	5.3	11.7	3.4	3.5
Average w_{final}		24.8	26.7	20.6	23.4	20.0	26.0	16.1	20.6	10.8	17.0	13.2	5.2	11.8	3.4	3.2
Test sample 6A				21.1	24.0			16.0	20.8	11.1	16.9	13.0	5.4	11.6		
6B		24.8	26.5	21.1				15.3	20.4	11.0	17.2	13.7	5.2	11.6		
Average w_{final}				21.1				15.6	20.6	11.0	17.0	13.4	5.3	11.6		

5

The scope for the use of the centrifugal model testing method for the determination of the influence on the stability of a waste-heap of different technological systems of waste-heap formation

Basic technological systems of waste-heap formation

The possibility of using the centrifugal modelling method to investigate the extent of the influence of the various technological systems of waste-heap formation on the stability of a waste-heap is interesting. In order to test all the methods of waste-heap formation we took as our basic systems those which are characteristic in respect of their possible influence on the stability of a waste-heap.

System I. A bulldozed waste-heap with vehicular transportation and deposition in horizontal layers. The structure of the body of the waste-heap is horizontally layered. In natural conditions this type of waste-heap is the most stable.

System II. An internal waste-heap, deposited by means of a transporter-dumping bridge, equipped with several trickle streams. Small loads or heaps are deposited, laid one on top of the other from below in such a way that the lower heaps serve as supporting prisms for the upper. The structure of the strata is with reverse inclination of the layers in relation to the slant of the embankment slope.

System III. A waste-heap deposited from the transportation cantilever of a bridge or waste-heap former without intermediate trickle streams. The soil is fed in a continuous stream with shuttle movements of the bridge or waste-heap former. The soil is deposited in thin inclined layers, formed as the soil rolls down the slope of the waste-heap. As a result of the continuous movement of the transporter along the brow of the slope the width of overspill is very small. The added layer is supported at its footing along a narrow strip. The structure of the body of the waste-heap is in the form of oblique layers. Judging by observations in the natural state this is the least stable type of waste-heap.

When modelling all three systems of waste-heap formation, in the process of acceleration of the centrifuge, the conditions of the geometry of the loading are continually changing and violate the constancy of the similarity of waste-heap formation. As yet we do not know how substantial is this departure from the prototype and we thus regard the tests carried out as only approximate.

The plan of the experiment

The plan of the experiment enabling us to determine by the centrifugal model testing method the extent of the influence on the stability of a waste-heap of the technological systems of waste-heap formation provided for the securing of comparative data from tests carried out in identical conditions of scale on models constructed by methods which reflect the fundamental features of the three technological systems considered above.

In the conditions of centrifugal model testing a continuous increase in the loadings and of the equivalents of the linear dimensions of the natural object occurs during the periods of acceleration of the centrifuge. In order to 'top-up' the models with additional material they are stopped and the additional material then added. As a basic condition in seeking a solution to the problem under investigation five cycles of runs (loadings) were adopted. After each topping-up the models were accelerated according to an identical graph, corresponding to a rate of waste-heap formation of $Q = 800$ m^3/h, with an increase in the scale of modelling to $n = 125$.

In the case when during the fifth run (loading) at rotations of the model corresponding to the scale $n = 125$ failure of the model does not occur, the loading is increased in accordance with the same principle until complete failure of the model is achieved.

The starting material for all three types of waste-heap was constant – Kerchi tertiary clay, ground in an air-dry condition to an aggregate size less than 0.5 cm and wetted after this to a nominal norm $w = 28.4\%$. The adopted rate of waste-heap formation was kept constant in all three cases at 800 m^3/h.

In the models the stable initial density of the waste-heap mass is well reproduced in two cases: with loose deposition ('from the shovel') and with dense deposition ('with ramming'). However in the case of deposition 'from the shovel', as a result of the dissimilar configuration of the models, the conditions of their initial density cannot be observed at all the various stages of loading. Hence to resolve the latter problem the method was adopted for shaping the models of depositing the soil in portions of predetermined weight, proportional to the cross-sectional areas of the parts of the model being deposited. As the basic overall weight of each model we took the values of the final weight of the models tested earlier in the second problem with analogous conditions of moisture content and productive rate of waste-heap formation. Thus, for the latter problem two extreme values of the initial density of the waste-heap mass were determined, δ, in tonnes/m³:

For loose deposition: ≈ 1.19–1.20.
For dense deposition: ≈ 1.46–1.50.

Finally, taking into account the significance of the character of the distribution of moisture in the waste-heap massif, the tests with the densely laid-down waste-heaps were repeated under a 'rain' (sprinkling) device. In this case the overall average moisture content of the soil mass is maintained the same as in the remaining tests ($w = 28.4\%$), but at the time of deposition in the model the moisture content of the soil is fixed at 1% lower, i.e. equal to 27.4%. The deficient moisture content (to bring it up to 28.4%) is restored in equal portions, applied in spray form to the open surfaces of the models before each topping-up addition in turn.

Taking into account the fact that the model conserves only some features or characteristics, regarded as the main ones, of the prototype, it is necessary to discuss the special features and conventional characteristics of each type of model built to correspond to each of the three technological systems of waste-heap formation.

Models of Type I (fig 59), corresponding to the first technological system of waste-heap formation, are deposited in horizontal layers in five equal portions, each weighing $0.2\,W$, where W is the total weight of the model (at the time of its collapse).

During centrifuging the layers consolidate and their settlement in different parts of the cross-sectional area is proportional to the depth of the layer at the given spot. Only those parts of the model retain their original initial position which are in contact with the sandy foundation and with the slopes of the waste-

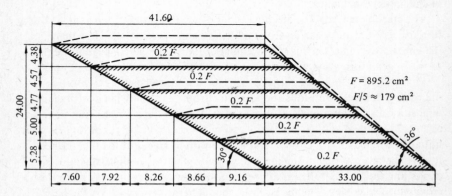

Fig. 59. Diagrammatic sketch of the models of Type I.

Scope: determination of influences on stability

heap of the preceding overspill, represented in the model by wooden wedges.

In accordance with the conditions of construction of the model, before each fresh acceleration (run) on the centrifuge one fifth of its total weight is laid in position. The models are built in accordance with predetermined weights and volumes. For this purpose the position of their immoveable supporting parts are recorded and also the position of the external slope, inclined at the angle of repose plus 1° to the horizontal (in our case $\alpha = 35° + 1° = 36°$). To obtain a horizontal surface of the waste-heap at the conclusion of loading each layer above the surface of the wooden wedges is deposited with a camber to zero, as shown by the dashed line in fig. 59.

Models of Type II (fig. 60), corresponding to the second technological system, are deposited as in the case of models of Type I in five portions of equal weight, each weighing 0.2 of the total weight of the model.

The soil mass is deposited, as shown in fig. 60 by the dashed line, with careful preservation of the condition of constancy of the contours of the supporting surfaces of the model. The initial height of the deposited heaps and the steepness of their internal slopes are derived magnitudes.

Models of Type III (fig. 61). In the third technological system of waste-heap formation the waste-heap is deposited in thin layers, inclined at the angle of repose to the horizontal.

Models of Types I and II are deposited on a 'consolidated slope' (imitated by wooden wedges), inclined at an angle of 30° to the horizontal. These conditions are also maintained for the models of Type III. Therefore to effect a transition from the angle of a 'consolidated slope' to the angle of repose a transitional wedge of part of the waste-heap soil weighing 0.115 *W* at an angle of 36° to the horizontal is deposited and loaded in an extra (zero) run on the centrifuge.

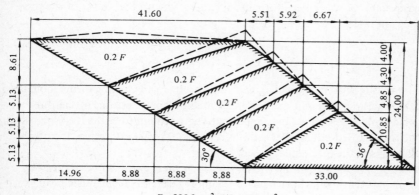

Fig. 60. Diagrammatic sketch of the models of Type II.

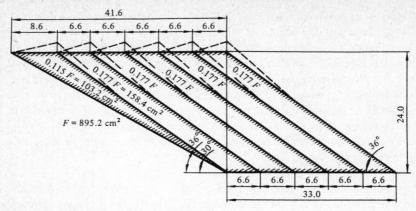

Fig. 61. Diagrammatic sketch of models of Type III.

The remaining soil, amounting to $1 - 0.115$ of the overall weight of the model, is deposited in the waste-heap in equal parts weighing $(1 - 0.115)W/5 = 0.177 W$ each. Deposition is carried out in layers, inclined at an angle $\alpha_{rep} + 1° = 35° + 1° = 36°$ to the horizontal. The width of the supporting strip along the footing of each layer is taken as equal to one fifth of the width of the supporting strip for models of Type I, $33/5 = 6.6$ cm. The height of the layers is a derived magnitude.

For the placement of the whole volume of each additional deposition in turn to the model it is permitted, when depositing the soil, to widen each layer in the upper levels at the expense of a change in the angle of the slope by a magnitude not exceeding the settlement during centrifugal loading. This magnitude is determined in the first loading (run). A composite plan of the experiment to determine the degree of the influence on the stability of a slope of the technological systems of waste-heap formation, combined with the results of the tests, is set out in table 31.

Results of the tests

By comparison with the preceding tests the results of the testing of the models prepared by three different methods reflecting the main features of the three technological systems of waste-heap formation turned out to be the least successful.

The tests carried out singly (without repetition) could not reflect the precise relationships in the problem as posed.

However, we succeeded in clarifying (even in these extremely curtailed experiments) that the influence of the technological systems of waste-heap formation examined on the stability of waste-heap soils is not as great as had been supposed. The results of the tests, integrated with the plan of the experiment, are set out in table 31. As the final comparative results we recorded the equivalent critical values of the height of the waste-heaps corresponding to the moments of their first movement and the collapse of the waste-heap.

Comparing them we can reach a conclusion concerning the general tendency towards a successive reduction in the stability of the waste-heaps from those of Type I, deposited horizontally, to those of Type II, with multi-stream deposition in steps from bottom to top, and finally to the least stable Type III waste-heaps, deposited in oblique layers. This is confirmed by the concurrence of the tendencies observed both in the models and in the natural state.

The exception in the sequence noted consists of the waste-heaps of Type II in the series of loosely deposited waste-heaps which showed, by comparison with waste-heaps of Type I, an increase in critical height of from 5 to 9%. For this series of models comparatively small variations in the critical heights are characteristic: from -11 to $+9\%$, which is within the limits of the accuracy of the experiment.

With dense deposition the tendency towards a reduction in stability in the indicated sequence of the models becomes more pronounced and the reduction in critical height amounts on average to 24% for Type II and 30% for Type III, by comparison with waste-heaps of Type I. The same sequence in the stages of the average reductions of 28% for Type II and 34% for Type III was also observed for the series of tests under the 'rain' device.

In the natural state we can observe an initial density of the waste-heap intermediate between the 'loose' and the 'dense'. If we take as guide-lines in a first approach the average values of the reductions in the stability of the waste-heaps between those revealed for loose and dense deposition, then by comparison with the bulldozed waste-heaps we must consider the critical height of waste-heaps deposited from a multi-stream cantilever bridge as 8–9% lower, and the critical height of bridge-formed and 'Absetz' waste-heaps, deposited with no intermediate trickle streams, as lower by 18%.

The models also reflected in some degree the different influence of conventional 'rain' on the waste-heaps deposited by the various methods. Whilst in the case of the bulldozed waste-heaps the effect of conventional rain, which raised the overall moisture content of the soil by 1%, was imperceptible, in the case of waste-heaps deposited from multi-stream cantilevers analogous 'rain' reduced the critical height of the waste-heaps by 5–9%, and in bridge-formed waste-heaps without intermediate trickle streams by 3–11%.

The numerical values quoted, reflecting the variations in the stability of waste-heaps deposited by different technological principles, were obtained on the basis of single tests without repetition and therefore cannot be regarded other than as guide-lines in this field, in which until now there had been no concept of the possibility of the relevant measurements.

In figs. 62, 63, 64 graphs are presented showing the readings from settlement indicators S.I. and displacement indicators D.I. at the end of the fifth runs

Centrifugal model testing of waste-heap embankments 122

Table 31 *The plan and results of the experiment to check the possibilities for determi*

Series of tests	Types and system of waste-heaps		Type I. Deposition in horizontal layers					Type I
	Predetermined		Model number	Measured		H_{crit} (metres)		Model
	w (%)	δ (t/m³)		w (%)	δ (t/m³)	At start of displacement	At time of collapse	numbe
Loose deposition	28.4	1.20	128	28.2	1.20	29.6 (100%)	33.0 (100%)	126
Dense deposition	28.4	1.50	132	28.5	1.50	39.9 (100%)	39.2 (100%)	124
'Rain'	27.4 + 1.0	1.50	133	27.1 + 1.0	1.49	36.7 (100%)	41.0 (100%)	130

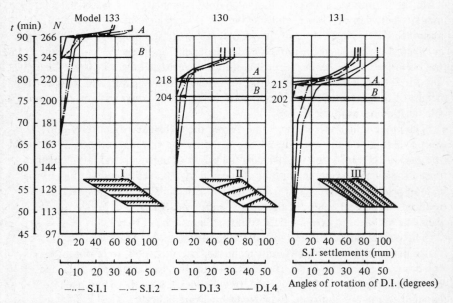

Fig. 62. Graphs of the deformations in time of loosely deposited models. t = time; N = number of revolutions per minute; 126, 128, 129, numbers of the models; I, II, III, the systems of waste-heap formation; A, collapse; B, commencement of displacement.

Scope: determination of influences on stability

influence of methods of deposition of a waste-heap on its stability (H_{crit})

position from a multi-stream cantilever				Type III. Deposition from a cantilever without intermediate trickle streams.				
asured		H_{crit} (metres)		Model number	Measured		H_{crit} (metres)	
%)	δ (t/m³)	At start of displacement	At time of collapse		w (%)	δ (t/m³)	At start of displacement	At time of collapse
1	1.20	32.5 (100%)	34.7 (100%)	129	28.2	1.18	29.4 (100%)	29.3 (100%)
4	1.44	27.8 (70%)	32.3 (82%)	125	28.5	1.42	27.6 (69%)	27.6 (71%)
9 .0	1.51	26.4 (72%)	29.3 (71%)	131	27.1 + 1.0	1.49	24.6 (67%)	26.8 (65%)

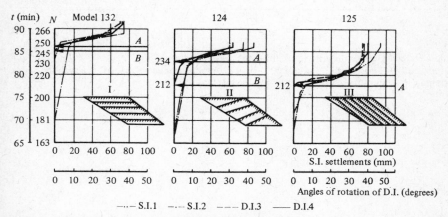

Fig. 63. Graphs of the deformations in time of densely deposited models. t = time; N = number of revolutions per minute; 124, 125, 132, model numbers; I, II, III, waste-heap formation systems; A, collapse; B, commencement of displacement.

(loadings) with the models, which recorded in time the processes of the collapse of the models of each of the three types with loose (see fig. 62), dense (see fig. 63) initial depositions of the materials, and also during the tests under the 'rain' device (see fig. 64). The graphs give some idea of the relative characteristics of the course of the collapses of each of the three types of models in time.

We can now give a positive answer to the question posed at the beginning of this section, but at the same time we must emphasise that in order to obtain reliable data the number of tests should be at least quadrupled.

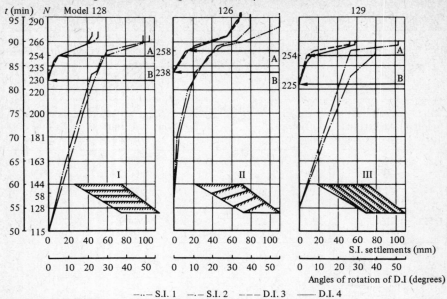

Fig. 64. Graphs of the deformations of models tested under the 'rain' device. t = time; N = number of revolutions per minute; 130, 131, 133, model numbers; I, II, III, the systems of waste-heap formation; A, collapse; B, commencement of displacement.

6

The stability of waste-heap embankments on weak, inclined bases

Basic principles

The stability of a waste-heap embankment, composed of non-saturable skeletal soils deposited at the angle of repose on a firm horizontal foundation, as expressed in terms of the critical height H_{crit}, is theoretically limited by that height at which the stresses arising from the self-weight loading of the waste-heap embankment exceed the strength of the soil aggregates comprising the waste-heap. At practically achievable heights in similar cases the waste-heap always remains stable unless it is subjected to additional dynamic actions: explosion, a seismic shock, the pressure from aqueous or air currents. The collapses of waste-heaps of skeletal soils observed in the natural state are connected, as a rule, with displacements over an inclined base or the squeezing out of weak soils from the base. In the areas covered beneath the waste-heaps the conditions of the bed of weak soils in the foundation can change in both the form and condition of the soils themselves and also in the conditions of their distribution. In such cases the deciding factors relevant to the stability of the waste-heap are:

1. The type and initial state (density, moisture content, strength) of the soils of the base.
2. The thickness z of the weak soils in the base which are capable of being squeezed out.
3. The angle of inclination relative to the horizontal β of the weak layers of the base and their location in relation to the direction of the fall of the slope of the waste-heap.
4. The type and initial state of the soils of the waste-heap.
5. The resultant (general) angle of the slope α of the waste-heap.
6. The productive rate Q of waste-heap formation or rate of advance of the waste-heap.

The factor method, described earlier, of experimenting with centrifugal models enables us to achieve a digital solution to a problem with three variables. Three factors were omitted from the quoted list of six deciding factors above, namely: by maintaining constant the form and initial state of the soils in the base and the waste-heap, and also the productive rate of waste-heap formation, the problem of the critical (limit) height of the waste-heap, given a sufficient number of experiments, can be resolved in the function of the three remaining variables:

$$H_{crit} = f(z, \beta, \alpha).$$

In accordance with the foregoing, by retaining in turn the constancy of two more variables, the extents of the influence of z and β on the critical height of the waste-heap and on the relationship $\alpha = f(H)$ can be determined.

Examples of the abbreviated determinations of the influence of z and β on the relationship $\alpha = f(H)$

Examples of the abbreviated determinations of the influence of z and β on the relationship $\alpha = f(H)$ were carried out in model waste-heaps composed of fine granitic ballast, deposited on weak foundations, laid down in the form of Tortonian clays and a mixture of quaternary sandy loams plus clayey loams considered earlier, the strength data-sheets for which were set out in figs. 20 and 21 and the Class II indices in table 6.

The soils of the foundation were assigned initial moisture contents, in percentage, of:

In the case of Tortonian clays: 38.

In the case of sandy loams and clayey loams 25.5.

The material for the model waste-heaps consisted of granite screenings with a coarseness of 1–5 mm and with initial moisture content $w_{init} = 8.5\%$, and characterised by the shear-resistance indices: $\varphi = 33°$ and $c = 0$. The type of models used, in connection with the introduction of the new variables z and β, was changed in appearance in accordance with fig. 65.

To ensure comparability of the results of modelling three additional conditions were adopted:

1. The effective radius of the model is defined as the distance between the axis of rotation of the model and the line of intersection of the initial surface of the base with the vertical plane passsing through the initial position of the brow of the waste-heap. In the calculations it is maintained at a constant $R_{eff} = 2.31$ m for the whole series of mutually comparable models. In other words, the initial position of the basic fixed point O (see fig. 65) in the model remains unaltered for all the models of the given series.

2. The initial height of the model waste-heap H_{heap}, defined in accordance with fig. 66 as the distance, along the vertical to the horizontal, from the brow or apex of the waste-heap to the point of intersection with the surface of the base (regardless of the inclination of the latter to the horizontal), is maintained in all the models of a given series as a constant (in our case equal to 24 cm).

3. The height of the slope in the model H_{slope} is defined as the difference between the points marking the brow and the toe of the slope (see fig. 66).

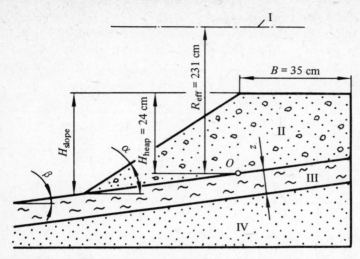

Fig. 65. Type of the model waste-heap on a weak, inclined base: I, axis of rotation; II, waste-heap; III, plastic soil; IV, sand.

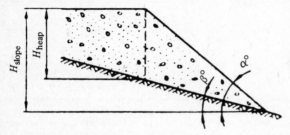

Fig. 66. Height of the slope of the waste-heap H_{slope} and height of the waste-heap H_{heap}.

As for the rest, the conditions of modelling and loading procedures are kept as before, based on an hourly rate of waste-heap formation of $Q = 800$ m³/h.

The boundaries of the variations of the variables were taken to be:

For z: $0.1\, H_{heap}$–$0.3\, H_{heap}$–$0.5\, H_{heap}$.

For β: $0°$–$5°$–$10°$.

For α: $27°$–$33°$.

The plan and sequence of the experiments are set out in table 32.

The results of the shortened experiments conducted, which bore an exploratory character, are presented in the form of graphs of the functional relationships $\alpha_{lim} = f(H)$ in fig. 67 for foundation soils of weak sandy loams plus clayey loams and in fig. 68 for weak, clayey foundation soils.

With the development in clays of continuous surfaces of slip the shear resistance of the soils in the foundation changes qualitatively. In fig. 68 the relationship $\alpha_{lim} = f(H)$ is represented by separate families of curves, corresponding to

Table 32 *Plan of the experiments to determine* $H_{crit} = f(z, \beta, \alpha_{init})$ *in model waste-heaps of strong skeletal soils on a weak base inclined at an angle* $\beta°$, *with a thickness of the weak stratum* $z = \frac{1}{10} i \times H_{slope}$ *and a waste-heap slope* $\alpha°_{init}$

Inclination of the base β (degrees)	Thickness of weak stratum, Z	$\frac{5}{10} H$ (12 cm)	$\frac{3}{10} H$ (7.2 cm)		$\frac{1}{10} H$ (2.4 cm)	
	Initial angle of slope α_{crit} (deg)	33	27	33	27	33
Initial angle of slope α_{crit} (deg)	Series	E (oblique)	C'	C''	D'	D''
0	27 A'	● 7–8	● 5–6		▲ 21–22	
0	33 A''	⊠ 1–2		⊙ 3–4		△ 19–20
5	33			☆ 9–10		
10	27 B'		◆ 13–14		▲ 17–18	
10	33 B''			⟡ 11–12		⟁ 15–16

Note: 1. The initial angle of the slope in models 7–8 was $\alpha_{init} = 24°$.
 2. The weak strata in the models were represented by (*a*) odd numbers – quaternary sandy loams plus clayey loams from Novy Razdol, where $w_{init} = 25.5\%$ and $\delta_{init} = 1.53$ t/m³; (*b*) even numbers – Novy Razdol tertiary clays, where $w_{init} = 38.0\%$ and $\delta_{init} = 1.21$ t/m³.

the conditions before and after the formation of continuous surfaces of slip.

For considerations of the convenience of measuring the models the basic curves of the graphs were constructed (heavy lines) for the relationship $\alpha_{lim} = f(H_{heap})$. Where the foundation soils are laid horizontally, when $\beta = 0°$, the identity $H_{slope} = H_{heap}$ occurs; in the remaining cases

$$H_{slope} = \frac{H_{heap}}{1 - (\tan\alpha/\tan\beta)}$$

The separate and the averaged curves, when converted according to this formula, are represented in the graphs in the form of the additional relationship $\alpha_{lim} = f(H_{slope})$. The graphs were constructed to correspond to the four series of experiments carried out.

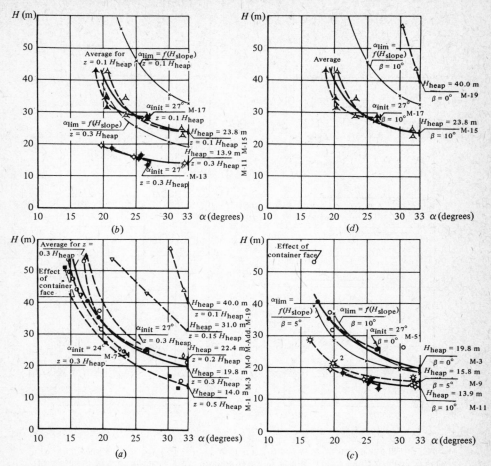

Fig. 67. The relationship between the function $\alpha_{lim} = f(H)$ and the thickness of the stratum z_{init} of weak sandy loams and clayey loams in the foundation of a waste-heap, and between that function and the angle of inclination of the foundation to the horizontal β, in degrees.
(a) Series B'' and B' ($\alpha_{lim} = f(z); \beta = 10°$). (b) Series C'' and C' ($\alpha_{lim} = f(\beta); z = 0.1 H_{heap}$). (c) = Series A'', A' and E ($\alpha_{lim} = f(z), \beta = 0°$). (d) Series C'' and C' ($\alpha_{lim} = f(\beta)$, $z = 0.3 H_{heap}$).

Series A'' and A' characterise the relationship between the functions $\alpha_{lim} = f(H)$ and the thickness of the weak layer z in the foundation, expressed in fractions of the height of the waste-heap H_{heap} with the foundation soils laid horizontally (where $\beta = 0°$). The series B'' and B' illustrate the same relationships for an inclined foundation, where $\beta = 10°$. The series C'' and C' characterise the relationship between the function $\alpha_{lim} = f(H)$ and the inclination of the weak stratum in the foundation of the waste-heap with a thickness of this stratum of $z = 0.3 H_{heap}$. The series D'' and D' reflect the same relationships but with thickness of the weak stratum equal to $z = 0.1 H_{heap}$.

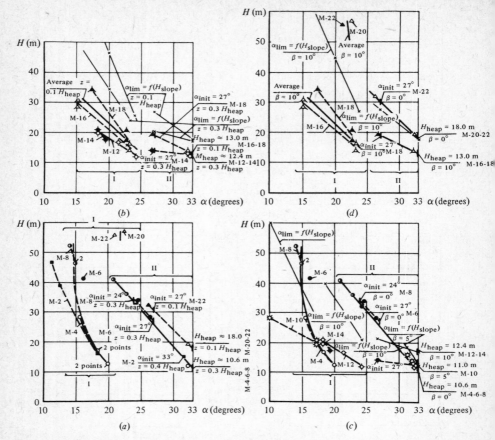

Fig. 68. The relationship between the function $\alpha_{lim} = f(H)$ and the thickness z_{init} of a weak layer of super saturated clay in the foundation of a waste-heap, and the angle of inclination of the foundation to the horizontal β, in degrees.
(a) Series B'' and B' ($\alpha_{lim} = f(z)$, $\beta = 10°$). (b) The series D'' and D' ($\alpha_{lim} = f(\beta)$, $z = 0.1 H_{heap}$). (c) Series A'', A' and E ($\alpha_{lim} = f(z)$, $\beta = 0°$). (d) Series C'' and C' ($\alpha_{lim} = f(\beta)$, $z = 0.3 H_{heap}$). I, after the development of a continuous surface of slip; II, before the development of a continuous surface of slip.

Characteristic of series C and D is the reduction, with an increase in the thickness of the weak stratum, in the influence of the angle of inclination of the foundation β (within the limits 0–10°) on the critical height of a stable waste-heap slope H_{slope}, which preserves an unchanging value of the resulting angle of the slope α.

The process of deformation of a weak foundation beneath a waste-heap embankment begins with gradual settlement of the foundation. When the deposition has been carried out in horizontal layers by the system for a bulldozed waste-heap, the magnitude of the settlement in the beginning remains proportional to the thickness of the deposited layer, i.e. beneath a horizontal waste-heap surface the settlement remains identical, but beneath the slope it reduces linearly and reaches zero at the toe (footing line) of the slope.

When tangential stresses are reached in the foundation soils which exceed their shear resistance, surfaces of shear develop either directly below the brow of the waste-heap or close to its projection. The shearing surfaces gradually spread upwards at an angle of $45 + \frac{1}{2}\varphi°$ to the horizontal.

In cases of rapid depositions on to weak foundations the shearing areas may spread along two arms at the same time, forming a wedge of active pressure implanting its point in the weak foundation and crumpling the adjacent sections on both sides of the foundation into folds. The lower section of the slope of the waste-heap which is transferring loads on to the foundation which do not exceed the load-bearing capacity of the foundation is moved outwards by the wedge of active pressure on the external side, at the same time crumpling the upper layers of the foundation into sliding folds (figs. 69 and 70).

With an increase in the strength of the clayey foundation the collapse of the slope of the waste-heap gradually changes in character and takes on the appearance of the well-known case of the development of a roundly-cylindrical surface of slip and rotation of the slope together with the foundation and accompanied by the formation of a protruding mound.

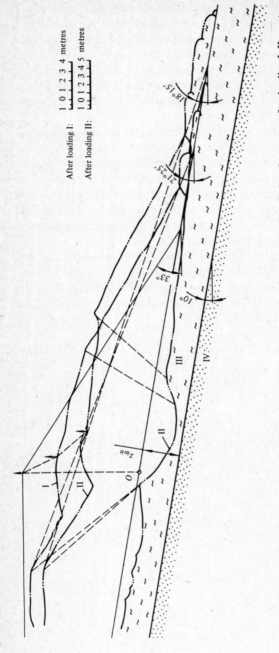

Fig. 69. Example of the collapse of a model ballast waste-heap on a weak (clayey) inclined foundation. I, contour after loading I; II, contour after loading II; III, soft, plastic clays; IV, sand.

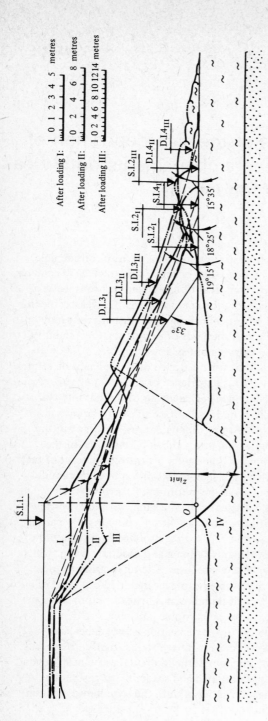

Fig. 70. Example of the collapse of a model ballast waste-heap on a weak (clayey) horizontal foundation. I, contour after loading I; II, III, the contours after loadings II and III; IV, soft, plastic clays; V, sand.

7

The conditions of the safe deposition of high waste-heap embankments onto weak (semi-liquid) foundations

Basic principles

The dumping areas of old shallow hydraulic waste-heaps occupy considerable areas of land. In the Kuznetsk Basin alone they occupy more than 2000 hectares, whilst nearby they continue to bury new areas of useful land beneath waste-heaps of 'dry' *semi-rock* overburden from the coal workings. The reason for this is the absence of satisfactory methods of depositing 'dry' waste-heaps safely onto hydraulic waste-heaps. There are neither methods of calculating their stability nor reliable, practical experience in carrying out such work.

Attempts to create a theoretical basis for the design of waste-heaps on semi-liquid old hydraulic waste-heaps (without experience of constructing such waste-heaps) have led certain authors to unreliable calculations of the stability (by applying the method of the roundly-cylindrical surface of slip to liquid soil masses) and to absurd conclusions, and to recommendations to carry out the waste-heap constructional work preserving intact the dried-out crust of the hydraulic waste-heap and conserving the liquid pulp beneath it. In point of fact, however, the deposition of waste-heaps of semi-rock soils on to old hydraulic waste-heaps, whatever the system of waste-heap formation, involves their intrusion into the liquid and plastic masses of the old pulp, with the overflow of a greater part of them from beneath the new waste-heap and with only partial trapping of the pulp beneath the waste-heap, with consolidation on account of the squeezing-out from the pulp of the gravitational and capillaric water, which passes into the porous body of the 'dry' waste-heap. The retardation of these processes by the suspension of a 'dry' waste-heap over the dried-out crust of a hydraulic waste-heap, in the presence of liquid masses beneath it, produces an increase in the potential threat of future catastrophic collapses.

On closer examination the problem of the deposition of waste-heaps of skeletal soils on hydraulic waste-heaps consists of two tasks: firstly, to prevent the squeezing-out and spreading of the pulp in the hydraulic waste-heap beyond the confines of the areas it originally occupied and secondly, to ensure the smooth immersion of the waste-heap of solid soil into the semi-liquid medium with no catastrophic situations.

The first task is resolved in a satisfactory way by using the *semi-rock* soils incorporating the waste-heap for the erection of a new, circular, protective dyke around the old hydraulic waste-heap whilst observing the condition that the height of the enclosure, throughout the whole time of the deposition of the waste-heap, should be maintained at datum points higher than the level of the semi-liquid pulp. Old alluvially-deposited dykes cannot serve as a reliable enclosure because of weakness. The new protective dykes should be located as much as possible lower than the alluvial dykes, and the latter should be regarded as part of the hydraulic waste-heap which is being covered over. The waste-heap should be deposited along the whole contour of the enclosure, with the pulp being squeezed from the contours towards the middle of the hydraulic waste-heap, or towards any adjoining elevations of the relief. The pulp thus squeezed out, and that which partially remains beneath the waste-heap, yields up its water comparatively quickly, the water passing into the porous body of the dry waste-heap. Since the latter consists of non-saturable hard soils, the excess water is filtered through them as through a seepage embankment and flows freely down the external slopes of the waste-heap enclosure without causing them any harm.

The second task is resolved by the selection of those cross-sectional profiles of the waste-heaps which preclude sharp collapses and sags beneath the mechanical devices working on the waste-heaps. When 'dry' waste-heaps are deposited on hydraulic lagoons in layers, the deposition of too high a layer should be avoided until the lower layers have completely embedded themselves into the pulp to the full depth previously established by experiment and corresponding to the predetermined elevation of the lower layers above the level of the pulp.

The height of the first layer, i.e. the pre-waste-heap layer can be set on the basis of two different principles, dictated by the boundaries of the minimum and maximum limits. The minimum height is defined by the necessity for the pre-waste-heap layer to stand clear of the water and out of the zone of capillaric wetting of the embankment from contact with the liquid pulp. As a rule, the elevation of the pre-waste-heap layer above the level of the liquid pulp H_{min} should be fixed at not less than 1 metre. And the maximum height of this layer above the level of the pulp H_{max} should be determined on the basis of the length of the jib of the waste-heap forming machine or drag line with which the 'dry' waste-heap is being deposited, and of the relationship $H_{max} = f(\alpha_{min})$, where α_{min} is the minimum critical angle of a stable slope after its collapse or slide.

In the first case, smooth movement of all the layers of the waste-heap and their embedding in the hydraulic waste-heap is achieved with any fixed limit to the critical height of the stages of settlement of the surfaces of the berms. This method is acceptable in a case where the new enclosing dyke of 'dry' non-saturable waste-heap soils cannot be raised to a great height at the start of the operations. An example of such a case is the Pritomsk waste-heap of the 'Krasnogor' mine in the Kuznetsk Basin coal fields, where on account of the proximity of a river it was not possible to lay down the base of the dyke straight away over the whole width and where it was necessary, in order to achieve this, to squeeze the hydraulic waste-heap away gradually, in the direction opposite to that of the river.

In the second case the advance of the waste-heap is associated with sharp, sometimes chaotic, collapses of the slope which cease completely by the time the slope has advanced over a distance somewhat less than the length of the boom of the waste-heap forming machine (drag line).

The second method is more economical, and it should be recommended in those cases when the use of the first method is not dictated by necessity. When using these methods it is necessary to take care that the soil is deposited on the brow of the waste-heap (advancing with the brow) to achieve the fullest possible expression of the pulp from beneath the embankment and to avoid excessive entrapment of the pulp between the separate positions of the soil being dumped. Finally, cases can arise when the raising of the enclosing dyke with maximum expression of the pulp will be carried out by the first method, but the subsequent stages of the waste-heap formation by the second method.

The critical stable profile should be selected individually for each waste-heap. It depends on the composition, condition and depth of the pulp of the hydraulic waste-heap, on the relief and character of its location, on the type of soils and height of the 'dry' waste-heap being erected, and also on the technological method of waste-heap formation. The problem is made more complicated by the presence over the pulp of a floating dried-out crust, especially if there is vegetation on it. If there is no crust the newly deposited non-saturable skeletal soils merge more smoothly into the viscous liquid and gradually form within it a certain stable submerged (semi-pulp) slope.

If we separate a strip of the crust of width AB, by slitting (fig. 71), and load it with a prism of a waste-heap $ABCD$, then this prism will sink into the pulp to a certain depth h.

In fig. 71 the following designations have been adopted:

$$T_1 = +Q \sin \alpha_{rep}; \quad T_2 = -P \sin \alpha_{rep};$$

$$N_1 = +Q \cos \alpha_{rep}; \quad N_2 = -P \cos \alpha_{rep};$$

$$U = \frac{h^2 \sin \alpha_{rep}}{2} \cdot \gamma_{pulp};$$

$$Q = F_{tot} \cdot \gamma_{heap}; \quad P = f_{water} \cdot \gamma_{pulp};$$

and φ_{heap} = angle of internal friction of waste-heap soils; φ_{pulp} = angle of internal friction of the pulp; F_{tot} = the total cross-sectional area of the prism; f_{water} = cross-sectional area of the part of the prism immersed in the water; γ_{heap} = bulk weight of the waste-heap soils; γ_{pulp} = bulk weight of the pulp; h = depth to which the prism sinks into the pulp.

Safe deposition onto weak foundations

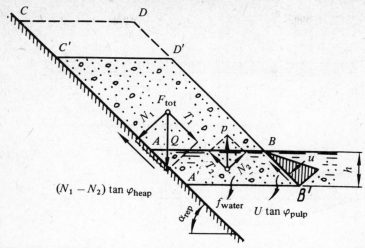

Fig. 71. Simplified diagram showing the immersion of the prism of the waste-heap into a viscous liquid possessing invariable viscosity.

If no difference existed in the retarding forces which arise as a result of the consolidation of the pulp on immersion of the prism, then the simplified scheme of the immersion of the prism, moving at the natural angle of repose of the soil α_{rep}, would correspond to the illustration in fig. 71.

In reality, however, as the waste-heap sinks into the viscous medium of the pulp the latter changes its physico-mechanical properties under the load on account of compaction and the relatively rapid expression of the water contained in it.

The initial shape of the vertical compressive loading, bearing on the surface of the viscous pulp from the layer of the 'dry' waste-heap, and the shapes of the corresponding compression deformations are illustrated in fig. 72. The protruding wedge of intrusion thus formed leads to partial entrapment of pulp between the succeeding depository loads. If the viscous medium or its crust has a load-bearing capacity of the order of σ_{load}, then the external wing of the intrusive wedge is displaced beneath the embankment slope at the spot where the normal (vertical) load from the weight of the column of waste-heap located above it exceeds the magnitude σ_{load}. When this happens, the upper part of the waste-heap and its slope collapses, leaving the lower part of the slope on the surface of the hydraulic waste-heap. As the wedge of the settling part of the waste-heap drops the lower part of the slope is displaced by this wedge in a horizontal direction (see fig. 72, b) and sometimes, together with the crust, forms a protruding bank. When this occurs, the angle of internal collapse of the waste-heap is defined by the magnitude $45° - \frac{1}{2}\varphi°$, where $\varphi°$ is the angle of internal friction of the waste-heap soils.

The angle formed between the external wing of the wedge of intrusion into

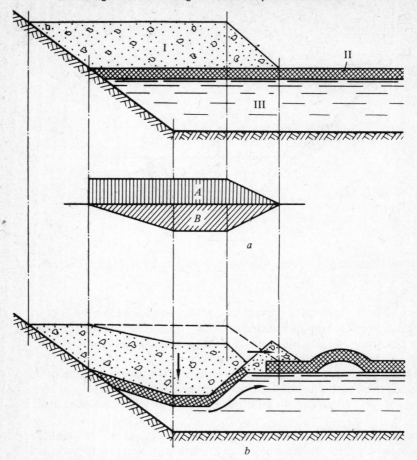

Fig. 72. The embedding of a 'dry' waste-heap into pulp in the presence of a crust.
A, Diagram of the initial vertical loading on the surface of the pulp; I, the waste-heap; II, the crust; III, the pulp. B, Diagram of the corresponding compression of the pulp; b, embedding of a wedge of the waste-heap into the pulp with the tearing away and horizontal displacement of the lower part of the slope.

the pulp and the horizontal is referred to by us as the angle of expression θ. This angle is not a constant index for a given viscous medium. It depends also on the degree of loading – on the weight of the prism of waste-heap soil being embedded into the pulp, on the initial depth of the pulp and on the relief of its bottom.

The presence of a floating crust and of a vegetative cover on the surface of old hydraulic waste-heaps creates additional, non-uniform (and therefore unsuitable for mathematical calculation) forces of resistance to the intrusion of 'dry' waste-heaps into the pulp of old hydraulic lagoons. Woody root systems stiffen the crust of the hydraulic waste-heap and confer on it the ability to

Safe deposition onto weak foundations

withstand at times considerable loads. However, with an increase in the load and the buckling of the crust in its lower levels the tensile stresses increase and lead to rupture of the crust and to catastrophic collapses of the soil masses deposited and suspended above it.

To avoid the suspension of waste-heaps and the associated growth of a potential threat of collapses, the root system of the vegetative cover and the dried-out crust of the hydraulic waste-heap should be cut through, before the deposition on them of 'dry' waste-heaps, to their total depth with deep slits disposed along and across the deposition front. Taking into account the scales of mining operations, the slits or gashes can be replaced by trenches and even by complete churning up of the crust.

To break up the continuity of the root system and of the crust one can employ the explosive method (with elongated charges), special machines (grubbing machines, trench-diggers, root cutters) and, finally, drag-lines with booms of sufficient length as to be able to operate directly from the deposited part of the waste-heap. Within the boundaries of the limits of the possible contact with liquid pulp and with the water squeezed from it only skeletal (draining) non-saturable soils should be allowed into the 'dry' waste-heap. Contamination by admixtures of weaker soils should not be permitted in an amount exceeding 10%.

Having adopted these basic principles it is essential (a) to establish the limit parameters of the waste-heap slopes which will ensure the safe conduct of operations on the given, actual hydraulic waste-heap and (b) to establish the conditions of the expression of pulp and the consolidation of its remaining part beneath the new waste-heap and, as a consequence of these processes, to determine the accompanying changes in the physico-mechanical properties of the pulp.

Yielding water under load, the pulp continuously and non-uniformly changes its moisture content, viscosity, and strength under the waste-heaps being deposited upon it. Consequently, the indices of these magnitudes, being unstable, cannot be included in calculations of the stability of a waste-heap on pulp. Hence the problems of the stability of a waste-heap and the condition of safe working on pulp are resolved experimentally. However, such experiments in the natural state are costly and unsafe. They can be reproduced with sufficient accuracy in centrifugal models constructed from the natural materials of pulp and 'dry' waste-heaps and subjected to bulk centrifugal loads which exceed the bulk weight loads of the natural state by as many times as the linear dimensions of the model are smaller than the dimensions of the natural object.

In the models we establish the relationships between the height H of the 'dry' waste-heap above the level of the pulp, the initial depth of the pulp z_{init} and the inclination of the bottom of the pulp $\beta°$, and such parameters of the waste-heap as the critical (minimum) angle of collapse of the slopes α_{min}, the angles of expression of the pulp θ, the depth to which the waste-heap sinks into the pulp

h, the width of the dangerous zone on the berms and working areas of the 'dry' waste-heap B.

In the models we can establish the conditions of the expression of the pulp and the consolidation of its residue beneath the waste-heap and, as a consequence of these processes, we also determine the accompanying changes in the physico-mechanical properties of the pulp.

The resolution in models of the conditions for the safe deposition of 'dry' waste-heaps of non-saturable soils on old hydraulic waste-heaps

The problems of selecting, in models, the optimum cross-sectional profile of 'dry' waste-heaps deposited on old hydraulic waste-heaps will be considered in the two examples of the cases of deposition of pre-waste-heap layers of minimum and maximum height.

In the first case the hydraulic waste-heap contained a semi-liquid pulp with an average depth of 6 metres. The dried-out crust on it had a thickness of about 0.5 m and was reinforced by the intertwined roots of dense scrub. The proximity of a river required the gradual squeezing of the pulp away from its bank and the erection of an enclosing filtration dyke as the pulp was removed. The waste-heap was formed by bulldozing.

In the second case the average depth of pulp in the hydraulic waste-heap was 20 m and in particular places reached 30 m, but its consistency was less fluid. At the banks the dried-out crust was of the same thickness as in the first case, but it had no vegetative cover. The central part of the hydraulic waste-heap was covered by a shallow layer of water. The relief of the locality made possible the raising of high protective dykes and a pre-waste-heap layer of maximum height, as determined by the method of deposition specially developed for this purpose – by the walking excavators ESH-10/60.

The models were constructed from materials of the average compositions of natural pulp and mixtures of overburden soils (sandstone and siltstone), selected *in situ* - at the hydraulic waste-heaps and at the overburden faces of the mines. The hard soils of the overburden were given a preliminary passage through a laboratory crushing machine and a sieve with mesh size $d = 10$ mm. At the time of deposition in the models the materials were brought to moisture contents equal to the average values observed in the natural state. The granulometric compositions of the materials for the models are set out in table 33. The bank slope, inclined at an angle of $30°$, was imitated either by an embankment of soils of the 'dry' waste-heap or by a wooden prism of triangular section, laid at the bottom of the container-bath, into which the pulp was poured to the pre-set depth. The datum for the initial surface of the pulp was taken as zero for all the subsequent vertical measurements.

In accordance with the conditions for its preparation to receive the 'dry' waste-heaps, the crust of the pulp, cut into rectangles measuring 6 × 3 m, was

Soil deposition onto weak foundations

Table 33 *Granulometric composition and moisture content of the soil materials of the models*

Fractions (mm)	In the conditions of hydraulic waste-heap Type I		In the conditions of hydraulic waste-heap Type II	
	Sandstone in the 'dry' waste-heap (%)	Hydraulic waste-heap pulp (%)	Mixture of sandstone and siltstone in the 'dry' waste-heap (%)	Hydraulic waste-heap pulp (%)
10.0–5.0	2.6	–	} 20.4	–
5.0–3.0	14.0	–		–
3.0–2.0	31.0	–	28.8	–
2.0–1.0	12.8	–	14.5	–
1.0–0.5	6.2	–	8.3	–
0.50–0.25	5.7	11.7	9.9	–
0.25–0.05	6.9	9.1	} 13.1	9.8
0.05–0.01	9.8	43.0		65.4
0.01–0.005	1.5	14.0	} 5.0	11.1
Less than 0.005	9.3	21.8		13.7
Moisture content w percent				
	8.6	43.7	8.6	36.0
		Upper limit of plasticity 30.5%		Upper limit of plasticity 33.7%
		Lower limit of plasticity 21.4%		Lower limit of plasticity 21.7%

represented in the models at the corresponding scales: by scraps of canvas on the hydraulic waste-heap Type I and by small leaves of blotting paper on the hydraulic waste-heap Type II.

In constrast to the determinations of the critical heights of the waste-heaps as described earlier, the modelling of the deposition of 'dry' waste-heaps on hydraulic waste-heaps is carried out at previously established scales of modelling. In each of its runs (loadings) the model is brought up to these scales, as up to limit scales. This permits mutual comparison of the data of all measurements of the model and its deformations in its external contours and in the contours of the boundaries of the embedding of the 'dry' waste-heap into the pulp. The latter are measured in narrow (2–3 cm) slits made in the models either after each run or after a fixed cycle of runs.

The schedules for the rotational runs of the models were established on the basis of calculations incorporating the following rates of productive waste-heap formation, in m³/h:

On the hydraulic waste-heap Type I: 700.
On the hydraulic waste-heap Type II: 200.

The scales of modelling were: in the first case $n_1 = 40$ and in the second $n_2 = 133$, but in some cases was also increased to 1:200. In order to preserve conditions of comparability of the results, in the determinations of R_{eff} the latter was taken as the distance from the axis of rotation of the models to the initial surfaces of the pulp. The conditions for safe working on the two types of hydraulic waste-heaps under consideration which we were investigating are different. When modelling, therefore, different technologies were employed.

In the first case it is necessary to ensure the safe movement of comparatively light machinery – a bulldozer and dump trucks – over horizontal or slightly inclined surfaces of the waste-heap under construction. On these surfaces the permissible unevennesses, including crevices and depressions, should not exceed limits of the order of 0.4–0.5 m to ensure the free passage over them of the levelling machines. In this case a search is made, using the models, for the method of waste-heap formation (whilst observing the conditions indicated) which ensures the quickest discharge on to the hydraulic waste-heap zones inundated by pulp, by means of a push by a bulldozer, of the minimum volumes of the 'dry' waste-heap. The model is constructed in this case by the deposition of the material of the 'dry' waste-heap before each run of the centrifuge in portions – overspills which encroach on the hydraulic waste-heap by the width of a strip, equivalent to 3 m in the natural state, at a layer height above the pulp of 1 metre. The screened berm is simultaneously topped up to the same height above the preceding deposits. The models and their foundations are measured before and after each run (loading). This process can be understood from the measurement profile of a pre-waste-heap layer in fig. 73.

When the surface of the continually supplemented berm remains after the sequential run (loading) at datum + 1 metre, i.e. when the intensive intrusion of the pre-waste-heap layer into the pulp ceases, one can begin (without interrupting the forward advance of the pre-waste-heap layer) the deposition of the second layer.

The limit of the safe width of the pre-waste-heap berm will be determined by its gradual contraction up to the moment when crevices and depressions arise on the working berm of the second layer which exceed in height the equivalent of the tolerance of 0.4–0.5 m in the natural state.

The dimensions of the berms of the succeeding layers are selected in the same way. In fig. 74 is shown an example of a profile selected in this way for a waste-heap dyke with a height of 40 m. The model (fig. 74) was originally constructed in a scale of 1 : 40 and only during the modelling of the deposition of the upper layer converted (by trimming the surface) to a scale of 1 : 200 for the use in the foundation of pulp consolidated by the loads of the preceding layers.

Towards the end of the testing of the model the complete information is collated on the changes which have occurred both at the surface and in the foundation of the waste-heap to achieve a stable profile. This information is supplemented by the recordings of automatic recording devices which have registered the course of the deformation of the models in time.

The experiment for the second type of hydraulic waste-heap was arranged on

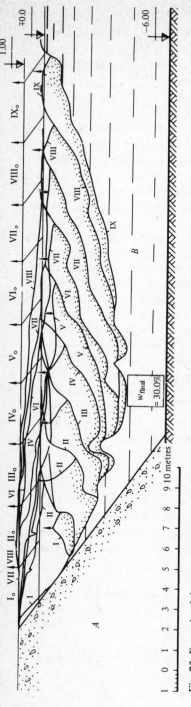

Fig. 73. Example of the measurement profile of a model 'dry' pre-waste-heap of Type I. Model no. 172/10B. w_{final} = final moisture content; A, the dyke; B, the pulp.

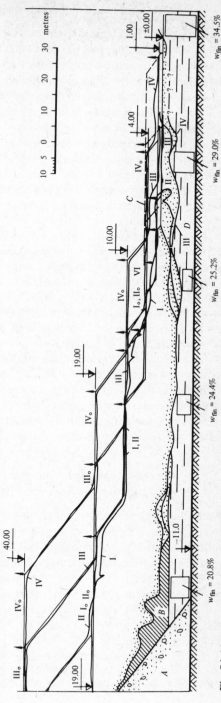

Fig. 74. Measurement profile of a model of a 'dry' waste-heap with H = 40 m on a hydraulic waste-heap of Type I. Model no. 173/11B. A, the dyke; B, section cut before run II; C, replenished by the amount of intrusion into the pulp; D, the pulp.

a wider scale to enable us to discover the laws governing the influence on the stability of a waste-heap of the additional factors: the width of the overspill, and the height of the waste-heap H above the level of the pulp, the initial depth of the pulp z_{init} and the angle of inclination of its bottom β.

An example of the measured profiles of a model of Type II is shown in fig. 75, where the measurements of the intermediate replenishments of the model are omitted and only the extreme positions of the profiles recorded: by dashed lines for the greatest spreads with minimum angles of slope and by heavy lines for the completed depositions, when the steepest angles were achieved by the slopes. The measurements were taken before and after each centrifuging, and excavations of the foundation were made only after the completion of a full cycle of deposition of each overspill (i.e. after several topping-up deposits within the confines of a given overspill).

These measurements enable us to establish the minimum angle α_{min} which a slope, deposited by a drag line, can achieve in periods of collapses, before the formation of a stable profile with a slope reduced to the angle of repose α_{rep} = 36°. By these same measurements we can also establish: the critical angle of expression of the pulp θ, the breadth of the danger zone B on the working berm, the maximum length of the sliding tongue L, reckoning from the foot of a fully deposited overspill the residue of unexpressed pulp beneath the waste-heap and the height of the terrace of protrusion of pulp in front of the completed overspill. The results of the measurements of numerous overspills in 17 models of waste-heaps of Type II are presented in tables 34 and 35.

In fig. 76 the survey measurements are presented of the actual collapse of a 'dry' waste-heap, raised up on the type of hydraulic waste-heap under consideration. Comparing it with the measurement profiles of the corresponding models we can observe the concurrence of the basic parameters of the slopes.

Changes in the physico-mechanical properties of pulp soils when subjected to loads from waste-heaps*

According to the granulometric classification of V. V. Okhotin the pulp from the hydraulic waste-heaps of both the types investigated belongs to the category of light and average clayey loam (containing in the region of 10-20% of particles measuring less than 0.005 mm), fairly dust-like and plastic.

We shall consider below only the case of a hydraulic waste-heap of Type II, whose pulp has an average initial moisture content of w_{init} = 36%, bulk weight γ = 1.82 g/cm^3, bulk weight of the skeleton δ = 1.34 g/cm^3 and coefficient of water saturation G = 0.97. Judging by the index of consistency b = 1.19, the pulp is in a fluid state.

During the process of model-testing, samples were taken of the pulp which

*This sub-section was written by Eng. A. P. Sakhno.

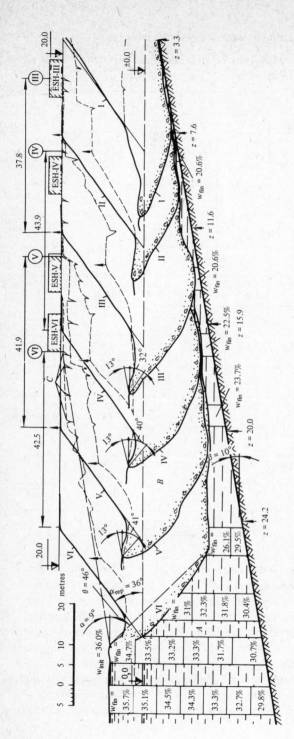

Fig. 75. Example of a measurement profile of a model pre-waste-heap of $H = 20$ m on a hydraulic waste-heap of Type II with an inclination of the bottom of $\beta = 10°$. Model no. 253/17. w_{init} = initial moisture content; w_{fin} = final moisture content; A, the pulp; B, the waste-heap; C, crevices.

Table 34 *Results of the tests of model waste-heaps deposited on a pulp of fine sediment with the average as denominator)*

Width of overspill (metres)	z (metres)	5						10
	H (metres)	5	10	15				20
		5	10	5+5	15	5+10	10+5	20
$\frac{6}{24}$	α_{min} (degrees)	15°–19° 17°	26°–32° 28°12'	26°	23°–28° 25°45'	18°	33°	26°–29° 27°
		–	–	–	–	–	–	13°–15° 14°
$\frac{6}{24}$	θ (degrees)	27°–36° 32°	21°–34° 28°36'	21°	22°–26° 23°45'	21°	22°30'	25°–38° 32°20'
		–	–	–	–	–	–	20°–20° 20°
$\frac{6}{24}$	B (metres)	5.9–6.9 6.2	5.0–6.5 5.9	6.0	5.3–12.0 8.7	7.8–16.8 11.6	3.0(?)	5.0–13 8.1
		–	–	–	–	–	–	25.3–2 26.5
$\frac{6}{24}$	L (metres)	1.0–1.6 1.4	1.0–2.3 1.4	0.5(?)	1.5–2.6 2.0	1.3	2.3	2.0–4. 3.3
		–	–	–	–	–	–	11.0–1 15.2
$\frac{6}{24}$	Remainder of pulp (%)	20–30 25	7–16 12	20–21 20.5	10–20 15	18–20 19	6–10 8	6.5–10 8.5
		–	–	–	–	–	–	9.0–36 22.5
$\frac{6}{24}$	Height of the terrace of protrusion (m)	0.6–1.6 1.1	?	1.5–2.0 1.7	?	3.0–4.7 3.8(?)	?	1.1–1. 1.2
		–	–	–	–	–	–	3.3–4. 3.8
Model numbers		1	4	2	6	3	5	15

remained behind beneath the model waste-heaps in order to determine their intermediate and final moisture contents. From table 36, which was compiled in accordance with the results of the determinations of the final moisture contents of the pulp, it follows that the height of the waste-heap H, the initial depth of the pulp z_{init} and, to a lesser extent, the inclination of the bottom of the pulp β all exert an influence on the reduction of the moisture content of the pulp. And, in addition, its consistency also changes in correspondence with the reduction in the moisture content of the pulp beneath the waste-heap. Even below a 10-metre waste-heap the final moisture content of the pulp is that which corresponds to a semi-solid state of soil, whilst beneath higher waste-heaps it corresponds to that

Soil deposition onto weak foundations

horizontal bottom of the Novobachat hydraulic waste-heap (showing values of limits

	20					30	
	10	20			40	20	40
+20	10	20	10+10	10+10+20	20+20	20	20+20
	13°–15° 14°	21°–23° 22°12'	32°	33°	–	20°–30° 25°30'	–
›	–	16°–18° 17°	–	–	32°	19°–25° 22°	32°
	41°–49° 45°52'	47°–53° 50°	50°	46°	–	57°–90° 74°15'	–
›	–	34°–35° 34°30'	–	–	38°	40°–73° 56°30.	40°
	18.0–27.2 23.9	10.9–22.6 15.9	6.0	3.0(?)	–	9.1–15.0 12.9	–
(?)		24.0–31.3 27.7	–	–	3.0(?)	28.6–30.6 29.6	3.0(?)
	3.1–5.9 4.3	4.7–8.9 6.0	4.7	6.4	–	2.7–10.7 5.8	–
.3	–	7.3–12.7 10.0	–	–	7.3	14.9–20.4 17.7	20.4
	13–20 16.5	6.7–10.0 8.3	14–17 15.5	8–13 10.5	–	4.5–15.5 10.0	–
–36.0 .0	–	7.0–24.6 15.8	–	–	2.0–11.0 7.5	5.0–16.0 10.5	4.5–13.5 9.0
	?	?	?	?	–	2.0(?)	–
	–	2.7(?)	–	–	2.7(?)	2.0(?)	2.0(?)
	7	10	8	9	11	13	14

of fully solid soil, and as such the pulp acquires considerable strength.

In parallel with the model testing of waste-heaps to determine the compressibility characteristics of the pulp, the pulp itself was subjected in its original state to compression tests in standard odometers using samples with an area of 60 cm^2. Loads were applied to the samples in steps of 0.5–1 kg and increased to 12 kg/cm^2.

The results of these tests are set out in fig. 77 in the form of the ratio of the modulus of settlement e, in mm/m, to the compressive forces σ, in kg/cm^2. For the same loads the corresponding values of the moisture content w, in per cent, and bulk weights γ, in kg/cm^3. are also set out in fig. 77. After the compression

Table 35 The results of tests on model waste-heaps of height $H = 20$ m deposited in overspills with a width of 24 m on a hydraulic waste-heap with different angles of inclination $\beta°$ of the bed beneath the pulp

β (degrees)	0			5			10		
z_{init} (m)	10.0	20.0	30.0	15.8	17.9	20.0	11.6	15.9	20.0	24.2
α_{min} (degrees)	13–15 14	16–18 17	19–25 22	13	13	13	13	13	13	9[a]
θ (degrees)	20–20 20	34–35 34°30'	40–73 56°30'	38	44	38	32	40	41	46
B (metres)	25.3–27.7 26.5	24.0–31.3 27.7	28.6–30.6 29.6	33.3	28.2	42.0	37.8	43.9	41.9	42.5
L (metres)	11.0–19.3 15.2	7.3–12.7 10.0	14.9–20.4 17.7	13.3	6.0	6.1	12.0	5.9	3.5	1.8
Pulp residue (%)	9.0–36.0 22.5	7.0–24.6 15.8	5.0–16.0 10.5	13.3–20.8 17.1	15.6–18.4 17.0	10.5–21.0 15.7	6.9–21.6 14.3	5.0–19.5 12.2	12.0–24.0 18.0	19.8–34.7[a] 27.2
Height of the terrace of protrusion (metres)	3.3–4.4 3.8	2.7(?)	2.0(?)	4.1	4.9	6.6	3.6	3.9	5.1	8.3[a]
Model numbers	15	10	13	18	18	18	17	17	17	17

[a] The indices were distorted by the proximity of the container wall.

Soil deposition onto weak foundations 149

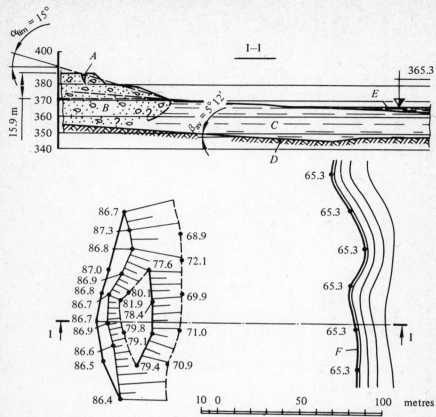

Fig. 76. Survey measurements of the collapse in the natural state of a 'dry' waste-heap, deposited on a hydraulic waste-heap of Type II. Landslide on self-formed waste-heap of no. 5 section of the 'Novobachat' mine (according to its condition on 11 July 1972). A, crevice; B, waste-heap; C, pulp; D, bottom of the hydraulic waste-heap; E, surface of water on 11 July 1972; F, retreat of water (drop in level).

tests the samples of the pulp soil were subjected to shear tests in the devices developed by N. N. Maslov with a shearing area of 40 cm^2. The results of these tests are also plotted in fig. 77. The angle of internal friction of the pulp soil, consolidated under different compressive loads, remained constant at $\varphi = 13°$, and the cohesion at $c \approx 0$.

By means of this ratio it is possible to characterise the strength of the pulp in its ultimate state beneath waste-heaps, for a pulp whose angle of shear resistance remains constant at $\varphi = 13°$. The moisture content and bulk weight of the samples taken from beneath model waste-heaps of varying heights lie almost exactly on the compression curves for $w = f(\sigma)$ and $\gamma = f(\sigma)$. Evidently, the residual pulp remaining beneath all the layers of the models achieves total consolidation.

Table 36 Results of measurements of the final moisture contents of the pulp beneath 'dry' waste-heaps

Inclination of the bottom of the pulp β (degrees)	Height of 'dry' waste-heap above level of pulp (metres)		Initial moisture content of pulp w_{init} (%)	Final moisture content of the pulp w_{fin} (%) beneath the waste-heaps, with an initial depth of pulp z_{init} (metres) of:							Reduction in moisture content $w_{init} - w_{fin}$ (%)
	Initial	Final		5.0	10.0	15.5	19.2	20.0	23.0	30.0	
0	10.0	27.3	36.0	21.7							14.1
0	15.0	19.5	36.0					21.9			14.3
0	20.0	29.1	36.0		21.5						14.5
0	20.0	38.6	36.0					20.4			15.6
0	20.0	48.5	36.0							19.6	16.4
0	40.0		36.0					18.8			17.2
10			36.0				22.5				12.3
10	17.3	34.1	36.0						23.7		13.5
10	18.3	32.6	36.0			20.6					15.4

Soil deposition onto weak foundations

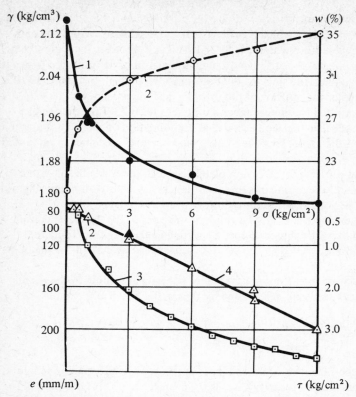

Fig. 77. The ratio of moisture content w, bulk weight γ, modulus of settlement e and shear resistance τ of the pulp to the compressive forces σ. 1, $w = f(\sigma)$; 2, $\gamma = f(\sigma)$; 3, $e = f(\sigma)$; 4, $\tau = f(\sigma)$.

During the process of the deposition of a 'dry' waste-heap there occurs the intrusion of the latter into the pulp at the expense of the heaving of the pulp and the settlement of its residue on account of its compression. In all the tests the thickness of the residual layer of pulp beneath the model waste-heap was measured and recorded.

By reference to the relationship $e = f(\sigma)$ it is possible to determine the settlement of the pulp as a result of its consolidation beneath loads σ, corresponding to the weight of the waste-heaps of varying height H with an initial depth of pulp z_{init}. To do this we first find the specific load σ corresponding to the moisture content of the pulp in its consolidated state w_{fin} on the graph of the relationship $w = f(\sigma)$. Next, using the relationship $e = f(\sigma)$ we determine the modulus of settlement e for the identified load σ. According to this modulus we then calculate the settlement for the initial depth of pulp. The difference between the total settlement of the base of the 'dry' waste-heap and the settlement on account of the consolidation of the pulp determines the magnitude of the outflow of pulp from beneath the waste-heap.

The calculations of the settlements and outflow of pulp for the models tested are given in table 37, from which it follows that the maximum outflow of pulp from beneath a waste-heap with a height of 10–20 m depends to a greater extent on the height of the waste-heap than on the initial depth of the pulp and depends practically not at all on the slope of the bottom of the pulp at gradients $\beta \leqslant 10°$. The average value of the outflow of pulp within the limits of the variables H, z_{init} and β investigated was determined in units z_{init} as $0.72\, z_{init}$, the average magnitudes of the settlements on account of consolidation of the pulp amounted to $0.20\, z_{init}$ and the average residue of pulp beneath 'dry' waste-heaps amounted to $0.08\, z_{init}$.

Outflows and settlements cease completely beneath pre-waste-heaps with a height of 20 m, and the raising of subsequent layers proceeds with practically no traceable deformations of the pulp residues beneath the waste-heap.

In fig. 78 an example is given of a graph of the ratio of the degree of consolidation θ of a sample of pulp (entrapped beneath a load) to time t. The initial height of the sample was $h_{init} = 2.5$ cm; the initial moisture content of the pulp was $w_{init} = 43\%$.

The conditions of the filtration of water in the pulp according to the results of the compression tests are determined by the coefficient of filtration

$$K_F = \frac{0.85\, ah^2\, \Delta_w}{4(1 + \Sigma_{av})t},$$

where a is the coefficient of compression, in cm^2/kg; Δ_w is the specific weight of the water (taken as equal to unity); Σ_{av} is the average coefficient of porosity for a given range of pressures; t is time, in seconds; h is the height of the sample, in cm, being tested beneath a pressure σ, in kg/cm^2.

The results of the calculation of the changes in the coefficient of filtration of the pulp under an increasing load are set out in table 38.

Conclusions and recommendations

The results of the investigations cited above enable us to make the following conclusions and recommendations.

1. The depth to which a 'dry' waste-heap sinks into the pulp is directly proportional to the height of the 'dry' waste-heap H above the surface of the pulp and the initial depth of the pulp z_{init}.

2. The depths to which 'dry' waste-heaps sink into the pulp at equal final heights of the waste-heaps, deposited in a single and in several layers are, as a rule, not identical; the higher the first layer (the pre-waste-heap) the greater are the depths of intrusion.

Table 37 *Calculation of the maximum outflow and consolidation of pulp beneath 'dry' waste-heaps according to the results of model testing and of compression tests on the pulp*

Determinations	Unit of measurement	Inclination of bottom of pulp $\beta = 0°$					$\beta = 10°$
		Model no. 6 $H = 15$ m $z_{init} = 5$ m	Model no. 7 $H = 10$ m $z_{init} = 20$ m	Model no. 15 $H = 20$ m $z_{init} = 10$ m	Model no. 10 $H = 20$ m $z_{init} = 20$ m	Model no. 13 $H = 20$ m $z_{init} = 30$ m	Model no. 17 $H = 18.3$ m $z_{init} = 15.5$ m
Total settlement of the base of the 'dry' waste-heap	Metres	4.5	17.3	9.1	18.6	28.5	14.2
Minimum thicknesses of the residual layer of pulp beneath the waste-heap	Metres, as per cent of z_{init}	0.5 10	2.7 13.5	0.9 9	1.4 7	1.5 5	1.2 7.8
Moisture content of the consolidated pulp	Per cent	21.7	21.9	21.5	20.4	19.6	20.6
Calculated compressive load (according to graph in fig. 77)	kg/cm²	5.1	4.9	5.5	7.2	9.0	6.8
Calculated modulus of settlement	mm/m	189	187	192	204	215	202
Settlement as a result of compaction (consolidation) of a layer of pulp z_{init}	Metres, as per cent of z_{init}	0.94 18.9	3.74 18.7	1.92 19.2	4.08 20.4	6.45 21.5	3.14 20.2
Outflow of pulp	Per cent of z_{init}	71.1	67.8	71.8	72.6	73.5	72.0

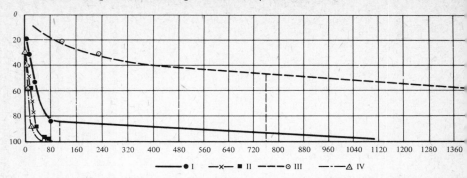

Fig. 78. Graph of the ratios of the degree of consolidation θ of a sample of pulp entrapped beneath a load to time t, in minutes. I, for the range of loads 0–0.5 kg/cm^2; II, for range 0–5.0 kg/cm^2; III, for range 7.0–8.0 kg/cm^2; IV, for range 0–9.0 kg/cm^2.

3. The speed at which a continuously deposited 'dry' waste-heap sinks into the pulp (as recorded during the model testing process by sensing devices and automatic recorders), up to the moment when the load-bearing capacity of the pulp is exceeded, increases uniformly; a sudden collapse of the embankment then occurs and afterwards intermittent, gradually abating immersion of the waste-heap with the expression of pulp from beneath it and, finally, when the expression of pulp ceases the sinking process comes to an end. The further, very slow and smooth settlement of the waste-heap occurs solely on account of consolidation – of compression with redistribution of water in the consolidating soils of the embankment, in the pulp residues trapped beneath the embankment and in the bottom of the pulp.

With an initial depth of pulp of up to 30 m the expression of pulp from beneath the waste-heap ceases completely when the embankment reaches a height of $H = 20$ m. Clearly, this height can be recommended as the rational height of the pre-waste-heap layer, since the raising on such a layer of the subsequent layers right up to a datum of +40 m will not be accompanied by any horizontal movements of the base. In such a case the conditions of stability of the slope do not require the laying down of a berm on the pre-waste-layer.

4. The depth to which a 'dry' waste-heap sinks into the pulp is inversely proportional to the width of the waste-heap forming overspill.

5. The process of immersion of a 'dry' waste-heap into pulp depends to a considerable extent on the dimensions of the sliding tongues of the 'dry' waste-heap which have pushed forward in the preceding overspill areas at the level of the surface of pulp and on the angle of expression θ.

The width of the creeping tongues L in the tests on models deposited in narrow 6-metre overspills varied within narrow limits (~1.5 to ~6 metres), increasing in direct proportion to the initial depth of the pulp z_{init}. The basic

Table 38 Determination of the coefficient of filtration K_F of the pulp according to the results of compression tests where $\gamma_{init} = 1.8\,g/cm^3$ and $w_{init} = 36\%$

Vertical compressive load σ (kg/cm²)	Coefficient of porosity — For a given load Σ	Coefficient of porosity — Average for a range of loads $\Sigma_{av} = \dfrac{\Sigma(n-1) + \Sigma_n}{2}$	Coefficient of compression $a = \dfrac{\Sigma(n-1) + \Sigma_n}{\sigma_n - \sigma_{n-1}}$	Height of soil sample under the load, $h = h_0 - \Delta h_n$, (cm)	Time t (sec)	Coefficient of filtration $K_F = \dfrac{0.85\,ah^2\,\Delta_w}{4(1 + \Sigma_{av})t}$
0	0.990					
0.5	0.811	0.900	0.357	2.276	4920	42×10^{-6}
0.5	0.811					
1.0	0.752	0.781	0.118	2.199	7200	95×10^{-7}
2.5	0.680					
3.0	0.665	0.672	0.030	2.112	73200	23×10^{-8}
5.0	0.616					
6.0	0.598	0.608	0.020	2.007	79200	13×10^{-8}
9.0	0.563					
10.0	0.556	0.559	0.007	1.953	157200	23×10^{-9}

factor affecting the magnitude of the advance of a creeping tongue is the width of the overspill. An increase in the width of the overspill to 24 m increases the width of the creeping tongues by 2-3 times.

6. The angles of expression θ in the case of shallow depths of pulp (of the order of 5 m) are in inverse proportion to the height of the waste-heap and will vary on average from $32°$ where $H = 5$ m to $21°$ where $H = 15$ m. At the same time, this relationship is almost unobserved in cases of a great initial depth of pulp. But with an increase in the initial depth from 10 metres to 30 metres, given identical values of the height of the 'dry' waste-heap $H = 20$ m, the angle of expression increases sharply (on average from 32 to $74°$, sometimes reaching $90°$ in the case of narrow overspills, and from $20°$ to $56°$ on average in the case of wide, 24-metre overspills).

In the case of a great initial depth of pulp a 'dry' waste-heap which is plunged steeply into the pulp experiences considerable resistance on the part of the abruptly counterposed pulp, in contrast to the unopposed sliding of waste-heaps over shallow, weak beds and areas of contact. This is also explained by the rapidly attained equilibrium of the 'dry' waste-heaps after immersion into the pulp, despite the incomplete consolidation of the pulp-base trapped beneath the waste-heap.

7. The effective resistance of the pulp also explains the increase, sometimes observed, in the steepness of the critical angles of the slopes, α_{min}, of 'dry' waste-heaps, composed of non-saturable soils, above the pulp, despite an increase in the height of the waste-heap H.

8. In accordance with tables 34 and 35 important data are established concerning the width B of the dangerous zones of crevices on the working areas being deposited. Although the width of the danger zones increases with an increase in the depth of the pulp, the deciding factor in establishing its maximum values is the width of the overspill.

9. The effect of the inclinations of the bottom of the pulp β on the indices α_{min}, θ, B and L can be seen from table 35. The width of the dangerous zone B increases with an increase in the angle of inclination of the bottom of the pulp. In a first approximation (in the conditions of the hydraulic waste-heap investigated) the width B is determined in terms of the height of the waste-heap H as:

$B = 1.6H$ where $\beta = 0°$,
$B = 2.1 H$ where $\beta = 5°$,
$B = 2.2 H$ where $\beta = 10°$.

The depth of immersion of the 'dry' waste-heap into the pulp at an inclination of the bottom up to $10°$ increases in proportion to the depth of the pulp to the same extent as in the case of a horizontal bottom, i.e. it is practically independent of the slopes of the bottom.

10. The composition of the pulp in the beach zone (where the pulp has been discharged from the pulp ducts) can differ considerably (although not always) from that of the pulp in the zone of fine sediment in the more distant sections of the hydraulic waste-heap. Comparative models should therefore be built of the extreme zones to establish the corrective coefficients when passing from recommendations given for one zone to the conditions of another zone.

11. The granting of permission for dumping operations on hydraulic waste-heaps demands individual designing of the surrounding dyke - in both the construction and the organisation of the dumping operations - as a special technological project. As basic planning and design starting data use should be made of survey measurements, geodesic and geological charts and selection of the boundary conditions of the stability of the waste-heaps by means of centrifugal model tests.

12. The opening of dumping operations must be preceded by: the construction of protective dykes, organisation of surface water drains and preparation of the base by the cutting of slits or complete grubbing up of the dried-out crust of the pulp.

13. For the dumping operations it is expedient to use the walking excavator ESH-10/60, which permits the work to be carried out as illustrated in fig. 79 with overspills with a width of 24 m at a height of the pre-waste-heap layer (above the level of the pulp) of $H = 20$ m.

The excavator bucket should be unloaded onto the brow of the waste-heap, maintained at the planned datum on the working surface. This ensures achievement of the maximum expression of pulp from beneath the waste-heap and avoidance of the entrapping of excessive volumes of pulp between the masses of the 'dry' waste-heaps. This method of deposition permits a gradual increase in the angle of the slope of the waste-heap in its upper part right up to the magnitude of the angle assumed by the free rolling of pieces of soil down a slope (35°-36°) whilst at the same time a sliding tongue is formed from the deposited soil in the lower part of the slope.

14. Advance from a completed overspill to a new one should proceed only after cessation of intensive settlement of the waste-heap along the whole of the strip formed by the preceding overspill.

15. The process of the expression of pulp from beneath the 'dry' waste-heap can be considerably accelerated by the use of camouflet explosions. For this the explosive charges should be laid along the front of the waste-heap at distances equal to the limit of the forward advance of the sliding tongue of the waste-heap in bore-holes in such a way that the base of the charge is at the level of

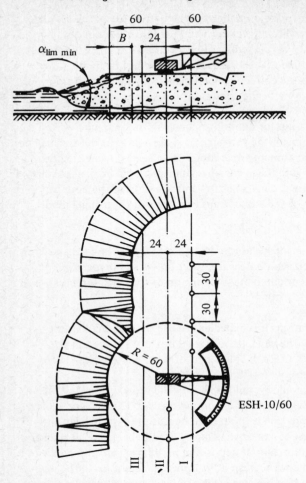

Fig. 79. Scheme of deposition of a 'dry' waste-heap of $H = 20$ m by drag line on a hydraulic waste-heap Type II. I, axis of first overspill; II, axis of second overspill; III, axis of third overspill.

the calculated position of the footing of the 'dry' waste-heap or the bottom of the pulp.

16. In 'dry' waste-heaps deposited on hydraulic waste-heaps constant observations of the rates and character of the settlements of the waste-heaps must be arranged so that in cases when danger-threatening crevices or excessive settlements develop the heavy equipment can be removed in good time from within their reach.

17. The raising of the level of the pulp as a consequence of its being expressed by the 'dry' waste-heaps should be compensated by a corresponding raising of the boundary datum points indicating the projected heights of the layers of the waste-heap, and also by a corresponding supplementary raising of the protective dykes in those places from which the waste-heap formation is not being carried out.

Making use of the recommendations cited above, the first mine in the Kuzbass (the 'Krasnogor') deposited without accident, during the course of one year, on old hydraulic waste-heaps, more than one million cubic metres of solid, hard overburden without occupying any new land, and reduced the mileage of the dump-trucks by 1.5 times.

8

The conditions of the stability of high waste-heaps on firm, inclined bases

The task and the programme of the experiment

The main mass of the semi-rock sandstones and siltstones of the overburden of the 'Krasnogor' mine in the Kuzbass is worked and shifted by means of a transportless system using ESH-10/60 and ESH-15/90 excavators which are capable of working out terraces up to 50 m in height in a single pass. The height of the internal waste-heaps is limited by the conditions of their stability on the inclined base in the worked-out area. The critical stability of a waste-heap depends on the shear-resistance of either the soils in the foundation or the soils in the waste-heap.

In the first case the deciding factor at the 'Krasnogor' mine is the surfaces of weakness encountered in certain well-known sections in the superimposed layers of the firm rocks of the base, inclined at angles of 8–20° to the horizontal and represented by sandstones and siltstones. In the second case, fundamental for the given mine, a further crucial factor is the changing (in time and according to the height of the waste-heaps) shear-resistance of the soils in the waste-heaps, deposited on inclined but completely stable bases, represented by thick strata of sandstone and siltstone without any reduction of the angles of internal friction below 20°.

The scope of centrifugal model testing excludes the construction of representative models of the bases, complicated by thin layers of semi-rock soils alternating with water-bearing interlayers of coals and carbonaceous argilites, which function as zones of weakness or as a dry lubricant between the layers. These shallow interlayers would have to be reproduced in the models in thickness of the order of some hundredths of a millimetre, which cannot be achieved in practical terms, as also the retention in the interlayers of gravitational water during the development of a powerful field of force during centrifuging. Accordingly, the possibilities of the centrifugal modelling of the internal waste-heaps is limited in the present case by the conditions of siting them on a stable foundation.

We interpret the height of the waste-heap H_{heap} and the height of the waste-heap slope H_{slope} as before in accordance with fig. 66. The angle of natural repose of the soils in the mine waste-heaps will vary within the limits 35–37°.

Stability on firm, inclined bases

When manoeuvering the excavators on the waste-heaps in the mine the slopes of the settled waste-heaps were sometimes undercut at angles of 45-57° to a height of up to 30 m. Not a single observed case had been recorded of the collapse of such undercut slopes despite the fact that they were sometimes left for a very long time. In fig. 80 an example is shown of a cross-section with an undercut slope of a waste-heap which stood intact for more than a year.

These observations prompted investigation of the possibility of reducing the volumes involved in the re-excavation of the waste-heaps of the 'Krasnogor' mine by means of undercutting their slopes at angles steeper than the natural angle of repose of the soil.

The investigations were carried out by the centrifugal model testing method in models constructed from soils selected at the mine in the most frequently encountered combinations of the compositions of the waste-heaps, in percentage, with variations of from 7 to 14% in the initial moisture content.

	Composition A	*Composition B*	*Mixture no. 2*
Sandstone	–	100	60
Siltstone	100	–	21
Argilite	–	–	4
Coal	–	–	5
Clayey loam	–	–	10

The base of the models was formed by continuous, intact, flat slabs of the very weakest siltstone, inclined at an angle β to the horizontal equal to 0°, 10° and 20°.

The stability of the waste-heaps was subjected to test: with an initial angle of the slope equal to the angle of repose $\alpha_{rep} = 36°$, and after successive undercuttings of the slope at angles of 50° and 70°. The order in which the model tests were conducted was the reverse of the order of the difficulty of the conditions for the maintenance of the stability of the models. In this way the programme of projected tests was reduced to a minimum.

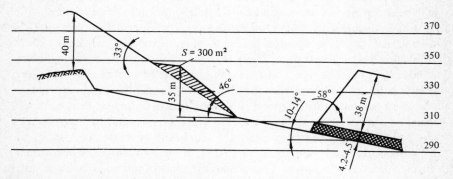

Fig. 80. Example of the natural-state dimensions of an undercut waste-heap slope which stood intact for more than a year. Section no. 2-B. Profile III.

For the construction of the models the hard soils of the waste-heaps were previously pulverised in a laboratory hammer mill and passed through a sieve with a mesh size of 10 mm. The physico-mechanical characteristics of the soils are set out in table 39.

Each type of soil composition was subjected to laboratory shear-resistance tests. The results of the tests, in the form of strength rating data are given in fig. 81. In addition, in the field the shear-resistance was checked of the superimposed layers which had fallen away in flakes on to the ground in the worked-out area. Their rating data are also shown in fig. 81. No angle of internal friction φ below 20° was discovered in either the first or second group of tests.

The modelling experiments were carried out in models of the type illustrated in fig. 82 (the dimensions are given in table 40), with a fixed (invariable) position of the point O, relative to which the constant effective radius of the models, $R_{eff} = 2.15$ m, was determined. The loading schedule (runs) for the models was fixed on the basis of the average productive capacity of the excavator – waste-heap former of $Q = 700$ m^3/h.

The model waste-heaps with slopes undercut at angles of 50° and 70° were additionally loaded with a model of the circular base of the walking excavator, represented by a metal disc with $d = 21.5$ cm, creating whilst at rest on the surface of the model a specific load of 0.0148 kg/cm^2. The edge of the disc was situated at a distance of 6 cm from the brow of the slope of the model waste-heap. During the modelling process the specific load from the disc increases in correspondence with the increasing scale of modelling. At the moment of the collapse of the slope it reaches recordable values which have an individual separate value for each model.*

Results of the experiment

The limits of the possibilities for the modelling of internal waste-heaps restricted us to the solution of only two problems: examination of the critical heights of high waste-heaps on a previously known stable, uniform base; and determination of the permissible heights of the undercutting of the slopes of waste-heaps (which have settled) at angles exceeding the angles of repose of the waste-heap soils.

In all, 92 experiments were conducted. The results of these tests are set out in table 41. In each square in this table are shown the model number and the critical height of the prototypes found in each experiment: of the waste-heaps (in the numerator position) and of the slopes (denominator), and also the final moisture content of the soil in the model. The experimental data have been grouped together, in accordance with the plan of the experiment, according to the composition and initial moisture content of the soils w_{init}, and also according to the variable values of the angles of the slope of the base β and the initial

*In the models tested they varied within the limits 1–3 kg/cm^2.

Table 39 *The physico-mechanical characteristics of Class II starting soil materials for the models of the internal waste-heaps of the Krasnogor mine*

Characteristic	Unit of measurement	Average indices for the soils					
		Sandstone		Siltstone	Carbonaceous argilite	Coal	Clayey loam
		Light	Dark				
Granulometric composition in mm:							
Fractions >5.0	Per cent	2.6	0.4	1.7	—	—	—
5.0–3.0	Per cent	14.0	3.9	13.0	—	—	—
3.0–2.0	Per cent	31.2	40.8	21.0	35.0	—	—
2.0–1.0	Per cent	12.8	15.4	8.0	20.8	—	—
1.0–0.5	Per cent	6.2	12.8	2.9	3.7	—	—
0.5–0.25	Per cent	5.7	5.1	2.4	1.4	—	0.5
0.25–0.05	Per cent	6.9	1.7	1.9	4.6	—	16.8
0.05–0.01	Per cent	9.75	9.9	32.1	18.8	—	44.8
0.01–0.005	Per cent	1.5	2.4	2.8	4.7	—	20.0
<0.005	Per cent	9.35	7.6	14.2	11.0	—	17.9
Specific gravity	g/cm³	2.66	2.67	2.62	2.38	1.69	2.76
Plasticity							
Upper limit	Per cent	—	—	—	—	—	37.9
Lower limit	Per cent	—	—	—	—	—	19.3
Plasticity index	Per cent	—	—	—	—	—	18.6
Hygroscopic moisture content	Per cent	0.81	0.84	1.16	2.46	—	—

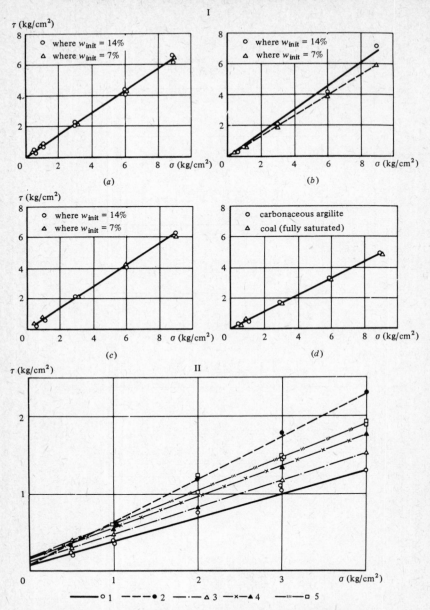

Fig. 81. Strength rating data for the soil materials of the model waste-heaps.
τ (kg/cm^2) = shearing force; σ(kg/cm^2) = normal load; I, laboratory tests: (a) sandstone 100% ($\varphi = 36°$, $c = 0.1$ kg/cm^2) (b) siltstone 100% ($\varphi_{14\%} = 37°$, $c = 0.1$ kg/cm^2; $\varphi_{7\%} = 34\%$; $c = 0.1$ kg/cm^2); (c) mixture no. 2 ($\varphi = 35°$, $c = 0.1$ kg/cm^2); (d) $\varphi = 29°$, $c = 0.1$ kg/cm^2. II, field tests: 1, waste-heap mixture, copiously flooded with water at time of testing ($w = 16.2$–17.2%, $\varphi = 17°30'$; $c = 0.05$ kg/cm^2); 2, waste-heap mixture ($w = 8.3\%$; $\varphi = 29°20'$; $c = 0.05$ kg/cm^2), 3, argilite, copiously flooded with water at time of test ($w = 13.6$–14.4%; $\varphi = 20°00'$, $c = 0.1$ kg/cm^2); 4, argilite ($w = 11.2\%$, $\varphi = 23°00'$, $c = 0.1$ kg/cm^2); 5, siltstone ($w = 10.1$–13.1%, $\varphi = 23°40'$, $c = 0.15$ kg/cm^2).

Stability on firm, inclined bases

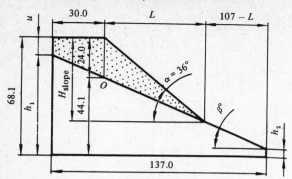

Fig. 82. Type and dimensions (in centimetres) of model.

Table 40   *Dimensions of the models (see fig. 82)*

Model type	β (degrees)	tan β	Dimensions (cm)				
			L	H_{slope}	h_1	h_2	u
I	20	0.364	66.3	48.1	55.0	5.0	13.1
II	15	0.268	52.5	38.2	52.1	15.6	16.0
III	10	0.1763	43.7	31.7	49.4	25.2	18.7

angles of the waste-heap slopes α before and after undercutting. By the conventional sign – a wavy line – we show the presence of an additional load in the form of the model of the base of the excavator.

Even in the first tests the need for further limitations of the modelling was established. The absence of clear and precise rules governing the relationship between the stability of the slopes and their soil compositions in the range of mixtures selected from those used at the mine was revealed. Within the limits of the accelerations created by the centrifuge it turned out to be impossible to achieve loads capable of producing the collapse of the slopes of the model waste-heaps, deposited in the form of all three soil compositions at the angles of repose $\alpha_{rep} = 36°$. In this case the comparatively wide range of strength of the given compositions of waste-heap soils is overlaid by variations in strength arising as a result of the accidental combinations of packing and interlocking of the skeletal aggregates during deposition of the model waste-heaps. Hence within the limits of the selected compositions of the waste-heap soils one can speak only of the overall minimum indices discovered in the process of testing all the models.

Moreover, we failed to produce either the collapse or the displacement over an inclined base of a single model with slopes deposited at the angle of repose even when the inclination of the base reached 20° and the height of the slope of the prototype waste-heap exceeded 90–100 m. These results correspond with the strength indices φ, in degrees, discovered when testing the starting materials for the models and their bases, which turned out to be not less than 20°.

Table 41 Composite table of the results of modelling the internal waste-heaps of the Krasnogor mine

Initial moisture content, w_{init} (per cent)	Angle of inclination of the base β (degrees)	Angle of the slope, α (degrees) Soils	7				14											
			36			50			70			36			50			70		
			Sa	Si	M-2	Sa	Si	M-2	Sa	Si	M-2	Sa	Si	M-2	Sa	Si	M-2	Sa	Si	M-2
0			No. 216/2 >90.5 w = 7%	No. 223/9 >87.5 w = 6.5%	No. 224/9 >95.5 w = 7.9%	No. 216/2 >28.1 w = 7%	No. 223/8 >26.5 w = 6.9%	No. 224/9 >29.5 w = 7.9%	No. 216/2 35.6 —	No. 223/1 19.1 —	No. 224/9 65.9 w = 7.5%	No. 204/31 >85.5 w = 8.6%	No. 225/10 >97.0 w = 8.1%	No. 207/34 >84.0 w = 13.0%	No. 205/32 87.0 w = 8.5%	No. 225/10 >101.0 w = 7.6%	No. 208/35 85.5 w = 12.1%	No. 206/33 42.5 —	No. 225/10 28.8 —	No. 209/36 34.4 w = 11.8%
10						No. 229/14 48.5/57.3 w = 7%	No. 230/15 >47.3/54.2 w = 6.7%	No. 228/13 >48.8/57.6 w = 7.6%	No. 229/14 18.3/19.6 w = 6.5%	No. 230/15 16.9/17.7 w = 6.6%	No. 228/13 28.0/29.8 w = 8%				No. 231/16 >51.1/60.0 w = 9.5%	No. 233/18 52.0/61.7 w = 8.1%	No. 234/17[a] 38.3/48.0 w = 12.3%	No. 231/16 45.4/49.4 w = 9.6%	No. 233/18 14.6/15.6 w = 8.3%	No. 234/17[a] 19.5/23.0 w = 12.2%
			No. 213/40 >37.9/86.7 w = 7.7%	No. 174/1 >54.2/112.6 w = 6.8%	No. 226/11 >47.2/97.0 w = 8%	No. 214/41 35.6/52.2 w = 7.1%	No. 222/7 49.9/78.8 w = 5.4%	No. 226/11 50.5/72.1 —	No. 215/42 14.2/16.7 w = 6.3%	No. 222/7 8.2/10.1 —	No. 226/11 35.9/41.3 w = 7.5%	No. 182/9 >47.7/99.6 w = 10.6%	No. 176/3 >50.5/104.0 w = 7.4%	No. 189/16 >47.6/99.5 w = 11.9%	No. 184/11 41.5/59.5 w = 10.3%	No. 178/5 >47.9/68.5 w = 9.3%	No. 191/18 22.6/31.2 w = 13.8%	No. 212/39 22.6/31.8 w = 8.2%	No. 180/7 20.3/23.4 w = 8.1%	No. 232/16 22.6/24.2 w = 13.4%
20			No. 217/1 >44.5/94.3 w = 6.3%	No. 222/7 >45.5/93.5 w = 6.3%	No. 175/2 >52.2/110.6 w = 7.1%	No. 217/1 32.4/46.5 —			No. 217/1 16.3/19.1 w = 7.3%			No. 199/26 >49.6/102.8 w = 8.3%	No. 177/4 >51.0/104.3 w = 9.9%	No. 190/17 >46.8/74.0 w = 11.9%	No. 185/12 43.0/79.5 w = 9.9%	No. 179/6 >47.6/68.5 w = 9.3%	No. 192/19 >52.0/75.6 w = 9.9%	No. 220/11 6.9/8.1 —	No. 180/7 23.8/27.5 —	No. 216/6 22.6/24.2 w = 12.7%
												No. 210/37 >52.0/105.3 —	No. 201/28 >49.0/101.5 —	No. 194/21 >44.8/95.6 w = 11.5%	No. 200/27 33.9/57.5 w = 8.3%	No. 202/29 >52.8/75.6 w = 8.7%	No. 193/20 >52.7/76.4 w = 9.7%	No. 227/12 24.7/28.4 —		No. 189/25 >27.0/31.9 w = 11.2%
												No. 220/5 >38.5/81.5 —	No. 220/5 >50.1/100.0 w = 9.3%	No. 211/38 >46.1/99.0 w = 11.9%	No. 195/22 >52.8/76.0 w = 8.7%	No. 227/12 >46.1/63.2 w = 8.7%	No. 196/23 30.3/44.0 w = 10.8%		No. 227/12 8.6	No. 227/12 8.6
												No. 183/10 >46.8/100.0 w = 10.2%		No. 221/6 >46.0/95.0 —	No. 220/5 45.0/65.6 —	No. 236/20 38.9/56.8 w = 8.7%	No. 197/6 52.0/76.8 —		No. 236/20 26.1/30.7 w = 8.7%	
																	No. 221/6 42.6/64.0 w = 12.7%			

Notes: 1. Sa = sandstone 100%; Si = siltstone 100%; M-2 = mixture no. 2.

2. The meanings of the entries in the squares are: model number, H_{heap}/H_{slope}, w = moisture content, ⌇⌇⌇ = with model of excavator.

Stability on firm, inclined bases

Thus the necessity to continue the tests in easier conditions of the stability of a waste-heap disappeared.

The worst case of a uniform base (i.e. having no weakened zones) corresponded to a solid unbroken thickness of siltstone which had an index $\varphi = 20°$. In this case, at an inclination of the base $\beta = 20°$ and an angle of slope $\alpha = 36°$, the reserve of the resistance of the waste-heap to displacement across the base was equal to unity.

Waste-heaps deposited at the angle of repose as non-cohesive loose masses of broken soil, regain a certain proportion of cohesiveness between the individual aggregates as they increase in size and stand undisturbed for a given time. A process of consolidation and strengthening of the waste-heap soils through the depth occurs. This strengthening makes it possible to impart to the waste-heap slope angles steeper than those formed during the process of deposition for some time, sometimes lasting years in fact.

The determination of the critical heights of the undercuts was achieved in the models in the following order. The models were deposited with slopes inclined at the angle of repose 36° and twice subjected to centrifuging, brought to a maximum of 300–310 revs/min. Before the second run on the centrifuge the settled and consolidated model was topped up to its full original profile.

Before the third run (loading) the slope of the model was cut at an angle of 50° in such a way that the brow of the new slope coincided, as far as possible, with the position of the old. To achieve this we simultaneously topped up the upper surface of the model to its original height and installed on it the model of the circular base of the excavator in the form of a steel disc. The third run (loading) was carried out up to the moment, recorded in time (by automatic recorders), of the collapse of the slope, or in those cases when no failure was induced – to the maximum revolutions possible, as in the first accelerations.

We next proceeded in analogous manner with the second undercutting of the slope at an angle of 70° and the model was subjected to a fourth run (or loading) to the moment of its collapse.

During the model-testing process we recorded:

(*a*) The number of revolutions per minute of the model at the moments of the commencement of collapse and achievement of maximum revolutions before starting the braking process.

(*b*) The magnitudes of the settlements of the upper surface of the models at the moment of the commencement of collapse, measured with the aid of settlement indicators located on the surface at a distance of 2.5 cm from the brow of the slope of the model.

(*c*) The cross-sectional profiles of the models before and after each run (loading).

(*d*) The character of the collapse (without signs of displacement along the base or with signs of sliding over the base).

(*e*) The final moisture contents of the waste-heap soils at the time of dismantling of the models.

The character of both kinds of failure of the models may be seen in the examples of the measurement profiles shown in figs. 83 and 84. In both cases the collapsed models have the appearance in their upper part of an exposed wall

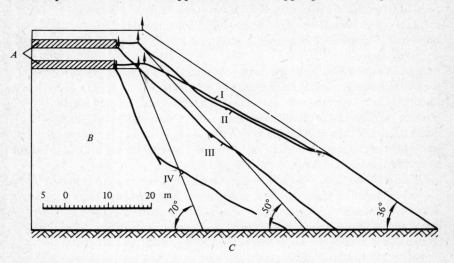

Fig. 83. Measurement profile of collapsed model no. 218/3 of a waste-heap (without signs of displacement along the base).
A, model of excavator; B, sandstone 100%, w_{init} = 14%; C, siltstone; I, II, III, IV, contours of the profile after runs I, II, III, IV respectively.

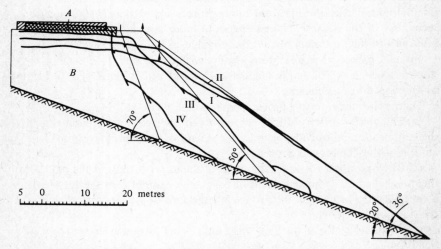

Fig. 84. Measurement profile of collapsed model no. 221/6 of a waste-heap (with signs of sliding over the base).
A, model of excavator; B, mixture no. 2 (sandstone 60%, siltstone 21%, argilite 4%, coal 5%, clayey loam 10%), w_{init} = 14%; I, II, III, IV, contours of the profile after runs I, II, III, IV respectively.

of rupture, and in their lower part the appearance of a scree disposed at the angle of repose of the soil, close to 36°

The height of the exposed parts of the walls of rupture and their inclination to the horizontal depend on the angle of the undercutting of the waste-heap slope. Thus, with undercuts at an angle of 50° the height of the exposed parts of the walls of rupture amounts to 17-34% of the initial height of the waste-heap, whilst the angles of their inclination to the horizontal will vary from 50° to 59°. With undercuts at an angle of 70° the corresponding height of the walls of rupture is 50-80% of the initial height of the waste-heap and the angles of inclination 60-69°.

The influence of the inclination of the base cannot be traced as being in conformity with any particular rule. With a horizontal base, in all cases except one the undercut slopes preserve their stability up to a greater height than where the bases are inclined. But at inclinations of the bases within the limits of only 10-20° the relationship between the stability of the undercut slopes and the inclination of the base is already lost.

To an even lesser extent can one trace the influence on the stability of an undercut waste-heap of the degrees of initial moisture content considered or of the difference between the three compositions of the soil masses examined. Thus, with an undercut slope at an angle of 50° the worst indices of stability revealed at an initial moisture content w_{init} = 7% by waste-heaps of sandstones were:

Where $\beta = 0°$: $H_{heap} > 81.1$ m.
Where $\beta = 10°$: $H_{heap} = 48.5$ m, $H_{slope} = 53.3$ m.
Where $\beta = 20°$: $H_{heap} = 32.4$ m, $H_{slope} = 46.5$ m.

With an initial moisture content of w_{init} = 14% the worst indices displayed by mixture no. 2 were:

Where $\beta = 0°$: $H_{heap} > 85.5$ m.
Where $\beta = 10°$: $H_{heap} = 26.6$ m, $H_{slope} = 31.2$ m.
Where $\beta = 20°$: $H_{heap} = 30.3$ m, $H_{slope} = 44.0$ m.

With undercutting of the slope at an angle of 70° the worst indices were revealed at an initial moisture content of 7% in waste-heaps of siltstones:

Where $\beta = 0°$: $H_{heap} = 19.1$ m.
Where $\beta = 10°$: $H_{heap} = 16.9$ m, $H_{slope} = 17.7$ m.
Where $\beta = 20°$: $H_{heap} = 8.7$ m, $H_{slope} = 10.1$ m.

With an initial moisture content of 14% the worst indices were obtained for waste-heaps of siltstones:

Where $\beta = 0°$: $H_{heap} = 28.8$ m.
Where $\beta = 10°$: $H_{heap} = 14.6$ m, $H_{slope} = 15.6$ m.

and for waste-heaps of sandstones:

Where $\beta = 20°$: $H_{heap} = 6.9$ m, $H_{slope} = 8.1$ m.

Given such a wide scatter of the indices of the critical height of the waste-heaps and the possibilities of continuous changes in the soil compositions within the limits of the six combinations of compositions considered and of their moisture contents we should be guided only by the lowest indices, without making any additional classifications either according to the inclination of the base or according to composition (within the confines of the three compositions considered) or according to initial moisture content (within the limits 7–14%).

That being so, the critical values of the height of the undercut slopes and waste-heaps will be determined by the magnitudes set out in table 42. However, these magnitudes also are to a certain extent accidental. For the determination of guaranteed minimum values of stable undercut slopes, with a probability $P = 0.99$ the cumulative curves shown in fig. 85 were constructed in accordance with the data in table 41 for the critical height of slopes undercut at angles of 50° and 70°.

With an undercut $\alpha_1 = 50°$ a stable slope can be guaranteed (with a probability of 0.99) at a height up to 28 m, and with an undercut $\alpha_2 = 70°$ up to 7 m.

Eng. A. P. Sakhno determined by back-calculations, by reference to the profiles of the collapsed models, the average values of the indices of the 'temporary stability' $\tan\varphi$ and c which occurred in the undercut models at the moments of their collapse (table 43). After reducing their values by the introduction of the coefficient of reserve (safety factor) $\eta = 1.1$ and inserting them into the corrective calculations, we determined the critical height of the undercut slopes with a pre-set reserve of stability. The results of these determinations are set out in table 44. They coincide with the guaranteed critical values of the height of stable slopes with a probability $P = 0.99$.

Practical conclusions

Despite the limited possibilities of the centrifugal model testing of waste-heap embankments of skeletal soils on semi-rock bases it does enable us to make a number of practical conclusions and recommendations. When making use of them, however, it should not be overlooked that the results obtained from the model testing and the recommendations given which are based on them relate only to the cases, guaranteed by preliminary bore-hole inspections, of the disposition of the internal waste-heaps of the 'Krasnogor' mine on stable

Table 42 *The minimum values of the height of waste-heaps and undercut slopes, in metres, as revealed by experiment*

Angles of inclination of the base β (degrees)	Angles of the slope α (degrees)					
	36		50		70	
	H_{heap}	H_{slope}	H_{heap}	H_{slope}	H_{heap}	H_{slope}
0	90.0	90.0	31.2	31.2	8.0	8.0
10	68.2	90.0	26.6	31.2	7.5	8.0
20	45.0	90.0	21.7	31.2	6.9	8.0

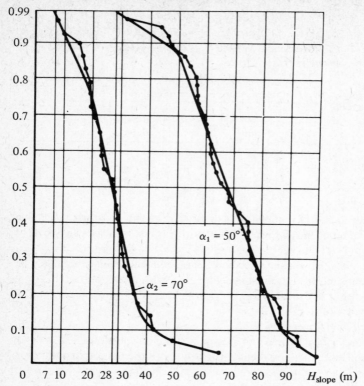

Fig. 85. Cumulative curves for the limit (critical) values of the height of a waste-heap slope, H_{slope}, undercut at angles $\alpha_1 = 50°$ and $\alpha_2 = 70°$; P = probability.

Table 43 *Average indices φ and c in (undercut) waste-heaps which have been allowed to stand undisturbed, as determined by back-calculations*

Soil	w_{init} (%)	φ	$\tan \phi$	c (t/m²)
Sandstone 100%	7	37°50′	0.775	1.63
Sandstone 100%	14	36°10′	0.728	2.38
Siltstone 100%	14	30°50′[a]	0.595	3.64
Mixture no. 2	14	37°10′	0.760	1.86

[a]The drop in the value of φ is explained by the fact that part of the surface of slip in the model came into contact with the base, for which $\varphi = 29°$.

Table 44 *Permissible critical heights of the waste-heaps considered H_{heap}, and of their slopes H_{slope}, in metres, on stable bases with undercut slopes at angles α, in degrees*

Angles of inclination of the base β (degrees)	Angles of undercutting of slope α (degrees)					
	36		50		70	
	H_{heap}	H_{slope}	H_{heap}	H_{slope}	H_{heap}	H_{slope}
0		60		28		7.3
10	45	60	24	28	6.8	7.3
18	33	60	20.5	28	6.4	7.3

foundations, composed of uniform beds of sandstones and siltstones, which have no zones of weakening in the form of interstratifications with soils and coals of lower strength.

The model testing results obtained and the recommendations given relate to waste-heaps deposited from soil masses consisting, with moisture contents of the waste-heap mixtures in the range 7–14%:

(*a*) entirely of sandstones;
(*b*) entirely of siltstones;
(*c*) of mixtures of sandstones and siltstones in any proportions;
(*d*) of the soil compositions already listed but with the permitted additions, in per cent of:

argilite: up to 4,
coal: up to 5,
clayey loam: up to 10.

The possibilities considered above of making use of the reserves of the stability of waste-heaps do not conflict with the currently effective 'unified safety regulations for the working of mineral deposits by the open-mining method', given the condition of an appropriate geotechnical basis and the limitations established individually for each actual specific case.

In respect of the conditions considered for the 'Krasnogor' mine these limitations and recommendations may be summarised as follows:

1. The minimum value of the angle of friction between the waste-heaps and the base is equal to 20°. In order to ensure a coefficient of reserve $\eta = 1.1$, the permissible angle of inclination of the base of the internal waste-heaps of the mine (excluding the known areas with zones of weakening in the ground) can be limited by the magnitude.

$$\beta_{max} = \text{arc tan}\, \frac{\tan 20°}{1.1} \approx 18°.*$$

*The Kuznetsk inspectorate of Grosgortechnadzor (state supervisory organization for mining technology) limited β_{max} to 15°.

Stability on firm, inclined bases

However, taking into account the possible accidental contamination of the surface of the base in the worked-out area and to avoid the sliding over it of waste-heaps at steep angles of fall of 12-18°, the base should be subjected to preliminary profiling by explosion according to the scheme shown in fig. 86, proposed by the Chief Engineer of the 'Krasnogor' mine, V. S. Vagorovsky with the object of creating a subsurface arresting stop.

2. On waste-heaps with the soil compositions considered, deposited at the angles of repose (of the order of 36°) on stable bases which have no weak interstratifications and contacts, temporary undercuttings of slopes may be permitted at an angle α not steeper than 45° with a limited height of the waste-heap H_{heap} and of the undercut slope H_{slope} corresponding to the graph presented in fig. 87.

3. At the approaches to the undercut slopes in the mine signs should be exhibited warning of the possibility of individual rocks and lumps of soil rolling down the slope.

4. Heavy mining equipment should be located outside the boundaries of a line drawn from the footing (toe) of the waste-heap upwards at the angle of repose of the order 36°.

5. The slope should be undercut by dragline, moving in reverse in compliance with the condition in paragraph 4 above.

In conformity with the recommendations enumerated above, over a period of two years undercutting of the slopes of internal waste-heaps was carried out at the 'Krasnogor' mine which made possible a reduction in the volume of re-excavated material by 4.2 million cubic metres.

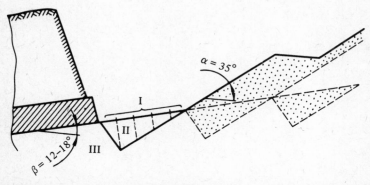

Fig. 86. Scheme for the profiling of the base of a waste-heap by explosion to create a subsurface arresting stop (according to V. S. Vagorovsky). I, slits; II, the body of soil blown up by explosion; III, the firm monolithic base.

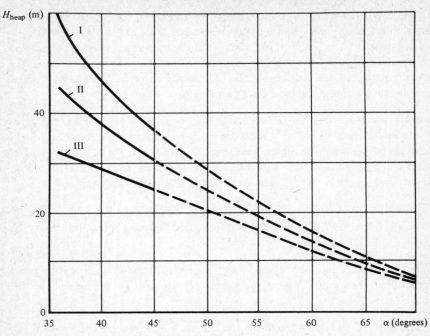

Fig. 87. Graphs showing the relationships between the permitted height of the waste-heap H_{heap} and the angle of undercutting of the slope α and the angle of inclination of a stable base β (for the 'Krasnogor' mine). I, where $\beta = 0°$; II, where $\beta = 10°$; III, where $\beta = 18°$.

9

Some recommendations regarding the use of the centrifugal model testing method in the solution of problems concerning the stability of waste-heap embankments

What may be, and what must not be resolved by the centrifugal model testing method

Model testing is a powerful means of investigating phenomena which are dependent on many variables. In particular, the centrifugal model testing method offers the possibility of finding answers to those problems of the stability of soil masses which do not lend themselves to other kinds of analysis. But the method also has its limitations. When posing problems it is essential to discriminate carefully between what can and what may not be obtained from this method.

In a model the similarity or identity is preserved of only the basic, characteristic aspects of the prototype which it represents. One must not demand from it total similarity with the natural object. The thought has already been expressed that demanding of a model total physical similarity with the prototype is equivalent to a denial of modelling. The observance of similarity in certain properties and phenomena can lead to divergences in others. It is important that the phenomena which are not taken into account should be immaterial and that inescapable errors when modelling should not be reflected in the accuracy of the determinations of the main phenomena under study.

We would utter a reminder concerning the erroneousness of the notion of the centrifugal model testing method as a method which automatically provides answers to any questions on the basis of single model tests. As has been established, in order to obtain a stable average result with an accuracy of 7–8% it is necessary to repeat an experiment four times.

The determination of the relationship between a final solution and a single variable requires the testing of a series of models whilst preserving the constancy of all except one of the variables in the construction and loading of the models. Stable conformities to a given rule or principle in serial tests allow us, in the absence of a substantial scatter of the final test results, to restrict the repetition of tests to two. Additive determinations of the directions of the rules or tendencies under study can be made in series of single (unrepeated) tests, but their accuracy drops sharply, as also in approximated solutions to problems.

The centrifugal model testing method is irreplaceable for the determination of the relationship between the stability of a soil waste-haep and individual, concrete variable factors and for factor experimentation with several simultaneously changing variables. But the method is cumbersome and labour-consuming. Its use is therefore inappropriate for the analysis of simple problems which are amenable to solution by simpler analytical methods.

The plan and quality of an experiment

The solution of every problem by the centrifugal model-testing method is preceded by the compilation of a plan of the experiment, which defines the aim, the starting resources and the sequence of conduct of the tests.

The plan should take into account the physical scope of the experiment, the assumptions adopted and the limitations, and also the degree of accuracy of the anticipated results. It should be precise and practical and take the form of a table, which permits the summary presentation of the basic principles of the experiment, and, by insertion in it of the results obtained, of the basic directions of the observed underlying laws also. The plan may meet the conditions for the finding of coupled or multiple relationships.

In the first case it is presented in the form of a simple flat grid, along the vertical and horizontal axes of which are located the appropriate values of the variables. At their intersections we plot the numbers of the models in order, their conventional designations and the magnitudes of the final solutions found as a result of conducting the experiment.

Plans for the study of multiple relationships, as a consequence of the sharply increasing volume of experiments, are as yet limited to triple relationships and can be presented either in the form of volumetric diagrams or in the form of a more complicated flat grid; for example, with additional oblique graphs, corresponding to the third variable.

The established plan of the experiment permits the conduct of the tests in any sequence. It is expedient, however, to make the extreme determinations at the beginning, with subsequent detailed determination of the intermediate values of the magnitudes being sought. This enables us to grasp the characteristic features of the results of the experiment in the very first few tests.

The tabular plans of an experiment should indicate those limitations which do not stem from the conditions given in a heading or the designations of the tabular graphs. All deviations from the plan which have occurred during its realisation should be accounted for and recorded for subsequent evaluation and incorporation of the appropriate correctives into the conclusions. Tests with substantial deviations from the norms established in the plan should be rejected and discarded after analysis of the reasons for these deviations.

Cleanliness in the preparation of an experiment and completeness of its

Recommendations: solution of problems

documentation are basic conditions, without the observance of which there can be no talk of comparability of the results obtained from the model testing either amongst themselves or between them and their prototypes.

Starting materials for model waste-heap embankments

The starting materials for model waste-heaps consist of the mineral soils being deposited in the prototype waste-heaps. As a rule, they are selected from a block of undisturbed overburden destined for the waste-heap. Depending on the conditions of previous experience the soil may be taken either in the form of undisturbed monolithic blocks (a core sample) preserved with paraffin, maintaining the soil structure and moisture content, or in the form of lumps and fines (in a heap).

In both cases the purity of the starting soil must be guaranteed free from extraneous contaminants both during the preparation of the models from the given soil in its pure form and in its mixture with other soils in the fixed proportions.

The selection of test samples for models from mixed waste-heaps should be excluded, since they yield accidental collections of soils which correspond neither to the average nor to the extreme compositions of the mixtures in the natural waste-heap.

The soil for the models undergoes a preliminary preparation according to pre-set conditions for the composition, grinding, mixing, wetting and leaving of the soil to stand before use.

Failure to observe the rules for the selection and uniform preparation of the soil for the models renders the results of tests with them non-comparable between theselves and with their prototypes. For comprehensive understanding and control of the quality and strength of the soils deposited in the models, each of their various types, and also the planned mixtures, are subjected to a shear-strength test and their components to determinations of the physico-mechanical indices of the granulometric composition, plasticity limit and specific gravity.

The documentation of the physico-mechanical properties and of the state of the soil materials of the models being tested is an indispensable part of the report data on the conduct of an experiment with models.

Concerning the prevention of certain errors which are possible when centrifugal model testing

During centrifuging the carriages with the containers and models turn around a horizontal axis O (fig. 88) at the points of their suspension from the balanced arm of the centrifuge through an angle $90° - w$, which is determined by the direction of the equally-acting (resultant) R from the combined vertical force of

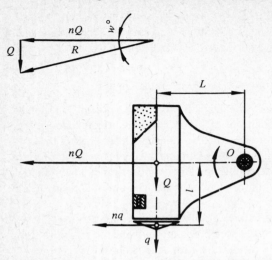

Fig. 88. Schematic diagrams of the loading of the centrifuge carriages.

the weight acting on the carriage Q with the horizontal centrifugal force nQ, where n is the scale of the accelerations (or of modelling).

The angle of inclination to the horizontal of the equally-acting (resultant) w is determined from the vector diagram (see table 45) as:

$$w = \arctan \frac{Q}{nQ} = \arctan \frac{1}{n}.$$

When modelling sloping models, situated (for convenience of installation of the measuring equipment) close to the ends of the containers which are lifted upwards during centrifuging, the angle of turn can be substantially reduced on account of the non-symmetrical loading of the container.

The possibility of the emergence of these errors when modelling with sloping models is eliminated by the four requirements used by G. Kh. Pronko.

1. The non-uniformity of the distribution of the load from the model in the container should be eliminated by the addition of weights on the opposite side with simultaneous balancing of the container on a lengthwise balance arm (fig. 89) using a level or plumb-bob.

Table 45   *The relationship between the angle w and the scale of modelling*

Scale of modelling	25	50	75	100	125	150	200	250	300
Angle w (rounded off)	2°20′	1°10′	45′	35′	30′	25′	20′	15′	10′

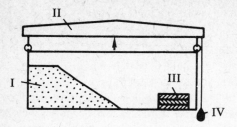

Fig. 89. Schematic diagram of the balancing of a container with a sloping model. I, the model; II, balance arm; II, the weights; IV, plumb-bob.

2. Removable counterweights q should be fixed on the end walls of the carriages which, during centrifuging, come to be below the horizontal axis of the rotation of the carriage Q (fig. 88).

The equilibrium position of the carriage during centrifuging is satisfied by the equilibrium of the moments

$$nQ \times O - (Q + q)L + nql = 0,$$

where L and l are arms of the forces acting relative to the axis O.

It follows that the necessary weight of the counterweight is

$$q = \frac{QL}{nl - L}.$$

Since the moment ql, which is restoring equilibrium, is dependent on the variable weight of the model which is included in the total load Q, it is desirable to have in the design and construction of the carriages of the centrifuge some means of remote control from the control desk over changes in the arm l by means of transpositions of the counterweight q beneath the bottom of the carriage along its axis.

3. The power supply for the apparatus for measuring the upward swing of the carriages and for recording them at the control panel should be arranged by use of storage batteries, in order to preserve constant voltage, i.e. it should be independent of the electrical supply system of the centrifuge.

4. The voltmeters used to record the angles of elevation of the carriages at the control-desk should be calibrated to an accuracy of 1° and in addition the position of the carriages when turned through a complete 90° should be recorded by light signals which light up at the control desk.

Recommendations regarding the use of the present book as a handbook on method

A handbook on method does not prescribe but shows by example how it is possible to solve a problem which is close in character to the given example.

In this book examples are cited of the use of the centrifugal modelling method to determine the extents of the influence on the stability of unconsolidated waste-heaps and bases of the main variable factors. The procedure developed for the organisation of the experiments and their processing is suited to these kinds of determinations. It can be employed for the solution of analogous problems in the planning, construction and operation of open-cast mines which have no analogues in the natural state with regard to the geological and technical mine-working conditions. By making use of our examples it is possible, at the stage of the preliminary geological survey of the minefields, to find one's bearings in the conditions of the stability of future waste-heaps, in relation to not only the character of the overburden soils, the height and configuration of the waste-heaps, but also to the technology and productive rate of waste-heap formation.

The procedure described for the modelling of waste-heaps should not be regarded as rigid instruction for any investigations of waste-heaps, but it will prove useful in various cases of centrifugal model testing. In the book examples are presented of the construction of plans of experiments and of the documentation of the tests carried out. They may not be repeated in their entirety in other work with models, if the plans for that work do not coincide with the plans for the experiments described earlier. They will, however, be useful for the development of new plans and of new forms of documentation in as far as it has been shown in these examples how this is done.

The present work has enabled us, simultaneously, in an initial approximation, to answer questions concerning the extent of the influence on the stability of unconsolidated waste-heaps of such factors as their soil composition, moisture content and initial density, the rate and method of waste-heap formation, and also to characterise the extent of the reduction in the stability of waste-heaps as a result of the development of continuous surfaces of slip in the body and the base of a waste-heap embankment during sliding displacements.

In two examples (clay and a mixture of sandy loams with clayey loams) in this work we have shown the character of the influence on the stability of waste-heap embankments of the main factors in their interconnection. There are insufficient completed experiments for generalising conclusions. But they have enabled us to clarify the character and tendencies of these relationships.

The practical application of the results obtained, in addition to their use at the coal and natural sulphur mines for which the investigations described were carried out, should find a place also for general, approximate determinations of: the limits on the height of waste-heaps when sliding displacements occur; the possibilities for mixed depositions in a waste-heap of sands and clays; the

Recommendations: solution of problems

limitations due to climatic conditions; the appropriate limits of the manipulation of the rate of waste-heap formation; the conditions of the deposition of 'dry' waste-heaps on weak bases; and the conditions of the exploitation of the reserves of stability of waste-heaps which have been allowed to stand and settle.

By the centrifugal model testing method the following results have been achieved.

1. It has been established that there is no unitary solution of the stability of an unconsolidated waste-heap without account being taken of five basic factors: the composition, moisture content and initial density of the soil masses in the waste-heap, and the rate and method of waste-heap formation.

2. A method has been found for determining the critical height of a waste-heap by the factor experiment method, as a simultaneous function of three variables: moisture content, initial density of the soils in the waste-heap and the rate of waste-heap formation with constant compositions of the soils and methods of waste-heap formation.

3. The extent of the influence of each of the five factors listed on the stability of unconsolidated waste-heaps has been determined in a first approximation.

4. The qualitative differences have been explained between the failures of clayey and sandy waste-heaps in cases where the critical height is exceeded, and in a first approximation we have established the extents of the reduction in the critical height of clayey waste-heaps at the moment of the formation in them of continuous surfaces of slip.

5. We have established the negative influence on the stability of a waste-heap of the initial consolidation of soft, moist soils in cases of the absence of drainage for the excess pore waters.

6. The conditions have been established of the positive and the negative influence on the stability of clayey waste-heaps of sandy additions of various moisture contents.

7. The conditions have been established of the safe deposition of waste-heaps on weak (semi-liquid) bases, making possible the use of the dumping areas of old hydraulic waste-heaps without occupying new areas of useful land.

8. We have established the conditions of the mobilisation of the reserves of stability of clearable internal waste-heaps in the interests of reducing the volumes involved in their re-excavation.

The aim of the present work will have been achieved if the quoted examples of methods and principles prove useful in the further solutions of problems concerning the stability of soil masses by the model testing method and if the results of the experiments described in it may serve to expand the concepts of the conditions of the stability of soil slopes and of the physical processes which occur when they fail.

Appendices

Appendix 1

Appendix 1 *Example of the determination of the basic parameters of the slopes of mo...*

Soil: Kerchi tertiary clays. Average initial moisture content $w = 27.29\%$. Average initial

Indices being recorded	Model numbers					
	6	7	16	17	Average	1.
Initial height of model (cm)	36.0 (where $t_I = 38'24''$)					3.

Conditions of commencement of failure of the model during loading (run) V

Number of revolutions of the model per minute	218	218	218	207		2
Effective radius (cm)	238	238	237	238		2
Scale of modelling	128.8	126.8	126.1	114.5		1.
Settlement of surface beneath indicator (cm)	5.5	4.8	4.6	5.3		
Height of model at brow (cm)	30.2	30.6	31.0	30.0		
Dimension of slope (cm)	48.1	45.1	47.9	45.7		
Average angle of slope	32°00′	34°10′	33°00′	33°20′	33°08′	3
Equivalent height of the waste-heap (metres)	38.3	38.8	39.2	34.4	37.7	

Final measurements of the model after loading V

Number of revolutions of the model per minute			218	
Effective radius (cm)	The same as at the commencement of failure of the models		238	The same as at commencement of failure of model
Scale of modelling			126.8	
Height of the model (cm) At the brow of escarpment			30.6	
At extreme crevice			–	
Dimension of slopes (cm) From brow of escarpment			68.7	
From extreme crevice			–	
Resulting angles of slope From brow of escarpment			24°00′	
From extreme crevice			–	
Equivalent height of waste-heap, (metres) At brow of escarpment			38.8	
At extreme crevice			–	

Appendix 1

e-heaps at the time of their failure

ity $\delta = 0.664$ t/m³. Run $t = $ min.

1	14	15	Average	8	9	12	13	Average	Notes
where $t_{II} = 32'00''$)				24.0 (where $t_{III} = 25'36''$)					
36	237	237		263	263	263	263		The values in bold type of the average of angle of the slope and the equivalent values of the height of the waste-heap correspond to conditions of collapse of the slope (formation of a continuous surface of slip)
41	241	241		242	242	243	243		
50	151.5	151.5		187	187	188	188		
4.4	4.0	4.4		2.7	2.3	2.9	3.0		
25.4	25.7	25.4		21.2	21.3	20.9	20.7		
38.5	37.1	39.0		32.0	31.9	31.9	31.6		
3°30′	34°40′	33°10′	33°32′	33°35′	33°40′	33°20′	**33°20′**	33°29′	
38.6	39.0	38.5	39.0	39.7	39.8	39.3	**38.9**	39.4	

37									
41		The same as at the commencement of failure of the models							
51.5									
26.0									
52.0									
6°32′									
39.4									

Appendix 2 Example of the establishment of the degree of accuracy of the determination of the critical height of a waste-heap

Selection of models	Number of variants (measurements) N	$\bar{x}$ Average arithmetical value of H_{crit} (metres)	Mean square deviation $\sigma_x = \sqrt{\frac{\sum(x_i - \bar{x})^2}{N-1}}$ (metres)	Coefficient of variation $v = \frac{\sigma_x}{\bar{x}} \times 100$ (%)	Maximum possible error ($P = 0.95$) $\sigma_{max} \approx 2\sigma_x$	Average error of arithmetical average $m_x = \pm \frac{\sigma_x}{\sqrt{N}}$ (metres)	Index of accuracy $p_x = \pm \frac{m_x}{\bar{x}} 100$ (%)
Models where $H_{init} = 24$ cm	7	39.21	$\sqrt{[(0.6^2 + 0.5^2 + 3 \times 0.1^2 + (-0.3)^2 + (-1.0)^2]/6}$ = ±0.54	1.38	1.08 m or 2.76%	±0.204	0.52
Models where $H_{init} = 30$ cm	7	38.73	$\sqrt{[(2.4^2 + 1.4^2 + 0.3^2 + (-0.1)^2 + (-0.2)^2 + (-1.3)^2 + (-2.5)^2]/6}$ = ±1.62	4.18	3.24 m or 8.36%	±0.613	1.58
Models where $H_{init} = 36$ cm	7	38.96	$\sqrt{[(1.5^2 + 0.2^2 + 0.1^2 + (-0.2)^2 + (-0.5)^2 + 2(-0.7)^2]/6}$ = ±0.77				
All models	21	38.96	$\sqrt{[(2.1^2 + 1.5^2 + 1.1^2 + 0.8^2 + 0.7^2 + 3 \times 0.3^2 + 0.2^2 + 0.1^2 + 0.0 + (-0.1)^2 + (-0.2)^2 + (-0.4)^2 + 2(-0.5)^2 + 2(-0.7)^2 + (-0.8)^2 + (-1.6)^2 + (-2.8)^2]/20}$ = ±1.05	2.70	2.10 m or 5.39%	±0.229	0.59

Appendix 3

Appendix 3 *Example of the verification of the significance of the variation in the average basic indices of the stability H_{crit} of sloping models of various height by the method of dispersion analysis*

Groups of models with initial height H_{init} (cm)		
24	30	36
X_i^{I}	X_i^{II}	X_i^{III}
39.8	41.1	40.5
39.7	40.0	39.2
39.3	39.0	39.1
39.3	38.6	38.8
39.3	38.5	38.5
38.9	37.4	38.3
38.2	36.2	38.3
$\bar{X}_1 = \dfrac{\Sigma X_i^{I}}{n_1} = 39.21$	$\bar{X}_2 = \dfrac{\Sigma X_i^{II}}{n_2} = 38.69$	$\bar{X}_3 = \dfrac{\Sigma X_i^{III}}{n_3} = 38.96$
	$\bar{\bar{X}} = \dfrac{\Sigma X_i}{N} = 38.95$	
$\bar{X}_1 - \bar{\bar{X}} = 0.26$	$\bar{X}_2 - \bar{\bar{X}} = -0.26$	$\bar{X}_3 - \bar{\bar{X}} = 0.01$
$(\bar{X} - \bar{\bar{X}})^2 = 0.0676$	$(\bar{X}_2 - \bar{\bar{X}})^2 = 0.0676$	$(\bar{X}_3 - \bar{\bar{X}})^2 = 0.00$
	$\Sigma(\bar{X}_i - \bar{\bar{X}})^2 = 0.1352$	
$\bar{X}_i^{I} - \bar{X}_1$	$X_i^{II} - \bar{X}_2$	$X_i^{III} - \bar{X}_3$
0.6	2.4	1.5
0.5	1.3	0.2
0.1	0.3	0.1
0.1	−0.1	−0.2
0.1	−0.2	−0.5
−0.3	−1.3	−0.7
−1.0	−2.5	−0.7
$(X_i^{I} - \bar{X}_1)^2$	$(X_i^{II} - \bar{X}_2)^2$	$(X_i^{III} - \bar{X}_3)^2$
0.36	5.76	2.25
0.25	1.69	0.04
0.01	0.09	0.01
0.01	0.01	0.04
0.01	0.04	0.25
0.09	1.69	0.49
1.00	6.25	0.49

1. The random dispersion is $D_z = \Sigma(X - \bar{X}_{i'})^2 = 20.83$.
2. The quantity of the degrees of the freedom of the random dispersion is

$$f_1 = N - m = 21 - 3 = 18;$$

where $m = 3$, the number of groups; $N = 21$, the overall number of variants, of measurements.

Appendix 3

3. The mean square of the random variation is

$$S_z^2 = \frac{D_z}{f_1} = \frac{20.83}{18} = 1.157.$$

4. The factorial dispersion is

$$D_x = \Sigma(\bar{X}_{i'} - \bar{\bar{X}})^2 \, n = 0.1352 \times 7 = 0.9464,$$

where $n = 7$ is the number of observations in one group.

5. The mean square between the groups is

$$S_x^2 = \frac{D_x}{f_2} = \frac{0.9464}{2} = 0.4732,$$

where $f_2 = m - 1 = 3 - 1 = 2$, the quantity of the degrees of freedom for the factorial variation.

6. The dispersion ratio is

$$F = \frac{S_x^2}{S_z^2} = \frac{0.4732}{1.157} = 0.409,$$

which is significantly lower than the critical value F_{crit} for the given number of degrees of freedom (2.18) - see tables in [23].

Where $\alpha = 0.05$: $F_{crit} = 3.555$,
Where $\alpha = 0.10$: $F_{crit} = 2.624$,
Where $\alpha = 0.25$: $F_{crit} = 1.499$.

Thus the observed differences in the average indices of H_{crit} are insubstantial.

Appendix 4 *Example of the use of the factor experiment method to determine the influence of the rate of deposition, moisture content and initial density of the soil materials on the stability of unconsolidated waste-heaps in centrifugal model testing*

(This example was completed with participation and procedural control of the statistical processing by Eng. A. A. Preobrazhensky.)

Calculation logbook

Tertiary (Tortonian) clays (Novy Razdol). Starting data, integrated in the plan of the experiment, are set out in table A.

1. Equation of the model of the process in a system of dimensionless variables

$$H_{\text{crit}} = b_1 + b_1 w' + b_2 \delta' + b_3 Q' + b_{12} w' \delta' + b_{13} w' Q'$$
$$+ b_{23} \delta' Q' + b_{123} w' \delta' Q',$$

where $b_0, b_1, b_2, \ldots, b_{123}$ are coefficients of regression, determined below (see paragraph 4); w' is the dimensionless (normalised) value of the moisture content of the material of the model; δ' is the dimensionless (normalised) value of the initial density of the material of the model; Q' is the dimensionless (normalised) value of the productive rate of waste-heap formation.

2. Formulae for the normalised variables

A. Moisture content

$$w'_i = \frac{w_i - w_0}{\Delta w},$$

where

$$w_0 = \frac{w_{\max} + w_{\min}}{2}; \quad \Delta w = \frac{w_{\max} - w_{\min}}{2},$$

units of the variation of moisture content.

B. Initial density

$$\delta'_i = \frac{\delta_i - \delta_0}{\Delta \delta},$$

where

Table A *Starting data, integrated with the plan of the experiment*

Number		Moisture content		Density		Rate of waste-heap formation		Critical height H_{crit} (m)	Calculation index
Of test	Of model	Actual w (%)	Normalised	Actual δ (g/cm^2)	Normalised	Actual Q (m^3/h)	Normalised		
1	34–35	28.2	−1	0.75	−1	800	−1	30.1	y_1
2	36–37	30.8	+1	0.75	−1	800	−1	21.5	y_2
3	54–55	28.3	−1	1.21	+1	800	−1	26.5	y_3
4	40–41	31.0	+1	1.20	+1	800	−1	15.0	y_4
5	52–53	28.3	−1	0.75	−1	2400	+1	27.3	y_5
6	44–45	31.0	+1	0.75	−1	2400	+1	21.8	y_6
7	50–51	28.7	−1	1.21	+1	2400	+1	24.3	y_7
8	48–49	30.8	+1	1.23	+1	2400	+1	18.2	y_8

Appendix 4

$$\delta_0 = \frac{\delta_{max} + \delta_{min}}{2}; \quad \Delta\delta = \frac{\delta_{max} - \delta_{min}}{2},$$

units of the variation of initial density.

C. Productive rate of waste-heap formation

$$Q'_i = \frac{Q_i - Q_0}{\Delta Q},$$

where

$$Q_0 = \frac{Q_{max} + Q_{min}}{2}; \quad \Delta Q = \frac{Q_{max} - Q_{min}}{2},$$

units of variation of the productive rate of waste-heap formation.

3. Computation of the calculation values of the dimension variables

A. Moisture content w

(a) $\quad w_{max} = \frac{\sum_1^4 w_{max\,i}}{4} = \frac{30.8 + 31.0 + 31.0 + 30.8}{4} = 30.90.$

(b) $\quad w_{min} = \frac{\sum_1^4 w_{min\,i}}{4} = \frac{28.2 + 28.3 + 28.3 + 28.7}{4} = 28.38.$

(c) $\quad w_0 = \frac{w_{max} + w_{min}}{2} = \frac{30.90 + 28.38}{2} = 29.64.$

(d) $\quad \Delta w = \frac{w_{max} - w_{min}}{2} = \frac{30.90 - 28.38}{2} = 1.26.$

Formula of normalisation of w:

$$w'_i = \frac{w_i - 29.64}{1.26}.$$

B. Density δ

(a) $\quad \delta_{max} = \frac{\sum_1^4 \delta_{max\,i}}{4} = \frac{1.21 + 1.20 + 1.21 + 1.23}{4} = 1.21.$

(b) $\quad \delta_{min} = \frac{\sum_1^4 \delta_{min\,i}}{4} = \frac{0.75 + 0.75 + 0.75 + 0.75}{4} = 0.75.$

(c) $\quad \delta_0 = \frac{\delta_{max} + \delta_{min}}{2} = \frac{1.21 + 0.75}{2} = 0.98.$

Appendix 4

(d) $\quad \Delta\delta = \dfrac{\delta_{max} - \delta_{min}}{2} = \dfrac{1.21 - 0.75}{2} = 0.23.$

Formula of normalisation of δ:

$$\delta'_i = \dfrac{\delta_i - 0.98}{0.23}.$$

C. Productive rate of waste-heap formation Q. In accordance with the conditions of the experiment $Q_{max} = 2400$ m³/h; $Q_{min} = 800$ m³/h. Consequently,

$$Q_0 = \dfrac{2400 + 800}{2} = 1600 \text{ m}^3/\text{h}; \quad \Delta Q = \dfrac{2400 - 800}{2} = 800 \text{ m}^3/\text{h}.$$

Formula of normalisation of Q:

$$Q'_i = \dfrac{Q_i - 1600}{800}.$$

4. Calculation of the coefficients of regression in a system of dimensionless variables

$$b_0 = \dfrac{y_1 + y_2 + y_3 + y_4 + y_5 + y_6 + y_7 + y_8}{8}$$

$$= \dfrac{30.1 + 21.5 + 26.5 + 15.0 + 27.3 + 21.8 + 24.3 + 18.2}{8} = 23.088.$$

$$b_1 = \dfrac{y_2 + y_4 + y_6 + y_8 - (y_1 + y_3 + y_5 + y_7)}{8}$$

$$= \dfrac{21.5 + 15.0 + 21.8 + 18.2 - (30.1 + 26.5 + 27.3 + 24.3)}{8}$$

$$= -3.962.$$

$$b_2 = \dfrac{y_3 + y_4 + y_7 + y_8 - (y_1 + y_2 + y_5 + y_6)}{8}$$

$$= \dfrac{26.5 + 15.0 + 24.3 + 18.2 - (30.1 + 21.5 + 27.3 + 21.8)}{8}$$

$$= -2.088.$$

$$b_3 = \dfrac{y_5 + y_6 + y_7 + y_8 - (y_1 + y_2 + y_3 + y_4)}{8}$$

$$= \dfrac{27.3 + 21.8 + 24.3 + 18.2 - (30.1 + 21.5 + 26.5 + 15.0)}{8} = 0.188.$$

$$b_{12} = \frac{y_1+y_4+y_5+y_8-(y_2+y_3+y_6+y_7)}{8}$$

$$= \frac{30.1 + 15.0 + 27.3 + 18.2 - (21.5 + 26.5 + 21.8 + 24.3)}{8}$$

$$= -0.438.$$

$$b_{23} = \frac{y_1+y_2+y_7+y_8-(y_3+y_4+y_5+y_6)}{8}$$

$$= \frac{30.1 + 21.5 + 24.3 + 18.2 - (26.5 + 15.0 + 27.3 + 21.8)}{8} = 0.438.$$

$$b_{13} = \frac{y_1+y_3+y_6+y_8-(y_2+y_4+y_5+y_7)}{8}$$

$$= \frac{30.1 + 26.5 + 21.8 + 18.2 - (21.5 + 15.0 + 27.3 + 24.3)}{8} = 1.062.$$

$$b_{123} = \frac{y_2+y_3+y_5+y_8-(y_1+y_4+y_6+y_7)}{8}$$

$$= \frac{21.5 + 26.5 + 27.3 + 18.2 - (30.1 + 15.0 + 21.8 + 24.3)}{8} = 0.288.$$

5. Construction of a full model of the process in a system of dimensionless variables

Using the computed values of the coefficients of the model (see paragraph 4) and substituting them in the equation of the model of the process (see paragraph 1) we shall obtain:

$$H_{\text{crit}} = 23.088 - 3.962w' - 2.088\delta' - 0.188Q' - 0.438w'\delta'$$
$$+ 0.438\delta'Q' + 1.062w'Q' + 0.288w'\delta'Q'.$$

The signs attached to the coefficients in this formula show a change in the value H_{crit} with a change in the corresponding parameter (if the coefficient is negative, then the height decreases, and vice versa), whilst the magnitude of the coefficient shows a change in the critical height in metres with a change in the corresponding parameter by a unit of variation (see paragraph 2).

6. Statistical analysis of the authenticity of the obtained coefficients of the model

On the basis of the tests described earlier (see appendix 2) the value was obtained of the mean square deviation σ_y in the measurement of H_{crit}, $\sigma_{H_{\text{crit}}} = 1.05$ m.

Appendix 4

This evaluation has 20 degrees of freedom ($f = 20$ for 21 measurements).

As applicable to the plan of the given experiment, in conformity with the theory of factor experiments [11] the mean square errors σ_{b_i} of the coefficients of the model are computed according to the formula

$$\sigma_{b_i} = 0.35\sigma_y = 0.35\sigma_{H_{crit}} = 0.35 \times 1.05 = 0.368 \text{ m},$$

since in our indexation y corresponds to H_{crit}.

Assuming a reliability of the evaluation of the coefficients of 0.95 (the risk of error in the determination of the authenticity of a coefficient is 0.05) and making use of the distribution tables of Student [18], we find that where $f = 20$ the Student ratio $t_{crit} = 2.09$. This being so, in accordance with the principles of mathematical statistics we can consider as reliable those coefficients of the model for which the condition is observed:

$$\frac{|b_i|}{\sigma_{b_i}} \geqslant t_{crit}.$$

In the opposite case the value of the coefficient is unreliable (it may be equal to zero) and it must be excluded from the model of the process.

The calculation of the statistical reliability of the obtained coefficients of the model is set out in table B.

Table B *Calculation of the statistical reliability of the obtained coefficients of the model*

Number of the tests	Index of the coefficient	Value of the coefficient as an absolute magnitude $|b_i|$	σ_{b_i}	$t = \dfrac{|b_i|}{\sigma_{b_i}}$	t_{crit}	Significance of the coefficient
1	b_0	23.088	0.368	62.76	2.09	+
2	b_1	3.962	0.368	10.77	2.09	+
3	b_2	2.088	0.368	5.67	2.09	+
4	b_3	0.188	0.368	0.51	2.09	−
5	b_{12}	0.438	0.368	1.19	2.09	−
6	b_{23}	0.438	0.368	1.19	2.09	−
7	b_{13}	1.062	0.368	2.89	2.09	+
8	b_{123}	0.288	0.368	0.78	2.09	−

Note: + denotes meaningful coefficients (essentially different from nought); − denotes unreliable coefficients (not essentially different from nought).

7. *Construction of an abbreviated model of the process*

By equating the non-positive coefficients with nought we can construct an 'abbreviated' model of the process:

$$H_{crit \text{ (calc)}} = 23.088 - 3.962w' - 2.088\delta' + 1.062w'Q',$$

where w', δ', Q' are the normalised values of the variables.

Appendix 4

Table C *Verification of the concurrence of the actual and calculated data when calculating according to the 'abbreviated' formula (with normalised variables)*

Test	H_{crit} (act)	H_{crit} (calc)	ΔH_{crit} = H_{crit} (act) − H_{crit} (calc)	Relative error, b (%)
1	30.1	30.20	−0.10	0.3
2	21.5	20.15	+1.35	6.3
3	26.5	26.02	+0.48	1.8
4	15.0	15.98	−0.98	6.5
5	27.3	28.08	−0.78	2.8
6	21.8	22.28	−0.48	2.2
7	24.3	23.90	+0.40	1.6
8	18.2	18.10	+0.10	0.5

Verification of the concurrence of the actual and the calculated data when calculating by the 'abbreviated' formula (with normalised variables) is shown in table C.

8. Conversion of the formula of the abbreviated model for the calculation of variables in the natural scale

By substituting in the formula of paragraph 7 the values of the magnitudes w', δ', Q' computed in paragraph 3 we find:

$$H_{crit} = 23.088 - 3.962 \left(\frac{w - 29.64}{1.26}\right) - 2.088 \left(\frac{\delta - 0.98}{0.23}\right)$$
$$+ 1.062 \left(\frac{w - 29.64}{1.26}\right) \left(\frac{Q - 1600}{800}\right).$$

After reduction of the similar terms the calculation formula takes the form

$$H_{crit\ (calc)} = 175.150 - 4.830w - 9.078\delta - 0.031Q + 0.001wQ \text{ metres.}$$

9. Verification and refinement of the calculation formula

By calculating the values of H_{crit} in accordance with the first variant of the formula obtained we can verify the concurrence of the actual and calculated values of H_{crit} (table D). Analysis of the data of table D shows that the calculated values of H_{crit} obtained are in all cases reduced by comparison with the actually observed values in the process of the experiment. This happens for the following reasons:

Appendix 4

Table D *Verification of the concurrence of the actual and calculated magnitudes of H_{crit} in accordance with the first variant of the calculation formula*

Test number	Conditions of experiment			Actual value $H_{crit\,(act)}$ (metres)	Calculated value $H_{crit\,(calc)}$ (metres)	ΔH_{crit} $= H_{crit\,(act)} - H_{crit\,(calc)}$
	w	δ	Q			
1	28.2	0.75	800	30.1	29.90	+0.20
2	30.8	0.75	800	21.5	19.42	+2.08
3	28.3	1.21	800	26.5	25.32	+1.18
4	31.0	1.20	800	15.0	14.53	+0.47
5	28.3	0.75	2400	27.3	25.17	+2.13
6	31.0	0.75	2400	21.8	18.61	+3.19
7	28.7	1.21	2400	24.3	20.02	+4.28
8	30.8	1.23	2400	18.2	14.74	+3.46

$$\Sigma \Delta H_{crit} = +16.99$$

(*a*) On account of a certain violation of the conditions of normalisation because of the impossibility of maintaining accurately the constancy of the magnitudes w_{max} and w_{min}, and also δ_{max}, δ_{min}, for all points of the experimental plan.

(*b*) Because of the inadequacy of the linear model of the process which was used to ensure true conformities to the rules which determine the relationship between the magnitudes of H_{crit} and the variables investigated.

In order to improve the concurrence of the actual and calculated data for H_{crit} we introduce a correction into the formula (see paragraph 8). The magnitude of this correction is

$$\overline{\Delta} H_{crit} = \frac{\Sigma \Delta H_{crit}}{n} = \frac{16.99}{8} = +2.124.$$

Taking account of the correction the calculation formula for H_{crit} takes the following form:

$$H_{crit\,(calc)} = 177.274 - 4.830w - 9.078\delta - 0.031Q + 0.001wQ.$$

The verification of the concurrence of the actual and calculated magnitudes of H_{crit} in accordance with the second, final variant of the calculation formula is set out in table E.

Table E *Verification of the concurrence of the actual and calculated magnitudes of H_{crit} when using the final variant of the calculation formula*

Test number	H_{crit} (act)	H_{crit} (calc)	ΔH_{crit} = H_{crit} (act) − H_{crit} (calc)	Relative error of calculation (%)	$(\Delta H_{crit})^2$	Mean square error in calculating H_{crit}
1	30.1	32.0	−1.9	6.3	3.61	$S_{H_{crit}} = \sqrt{\left(\dfrac{\Sigma(H_{crit}\,(act) - H_{crit}\,(calc))^2}{n-1}\right)}$
2	21.5	21.5	0	0	0	
3	26.5	27.4	−0.9	3.4	0.81	$= \sqrt{\left(\dfrac{15.05}{8-1}\right)} \approx 1.47$ m, or $\sim 6.4\%$
4	15.0	16.7	−1.7	11.3	2.89	
5	27.3	27.3	0	0	0	This will occur in 70% of the cases of calculations, but the maximum error to be expected is $\Delta_{max} = 2S_{H_{crit}}$ = 2.93 m, or $\sim 12.7\%$.
6	21.8	20.7	+1.1	5.0	1.21	
7	24.3	22.1	+2.2	9.1	4.84	
8	18.2	16.9	+1.3	7.1	1.69	
			−4.5 +4.6	Average 5.3%	Σ = 15.05	

Appendix 5 The M.Ts.M.-2 dynamometer for the remote measurement of compression stresses in models (developed by G. E. Lazebnik)

The measuring scheme and diametral cross-section of the M.Ts.M. dynamometer are shown in fig. 90, and a general view in fig. 91.

The device is a twin-membrane soil dynamometer, symmetrical relative to a mean plane, made of aluminium or magnesium alloys. As a result of a special way of connecting the strain resistances the readings of the M.Ts.M-2 dynamometer are independent of the direction of the acting force relative to its membranes. Therefore both membranes of the device are sensitive to pressure, the manifestation of a 'directional effect' is eliminated, and temperature compensation is ensured. The readings from the detector are recorded on standard strain-gauge instruments.

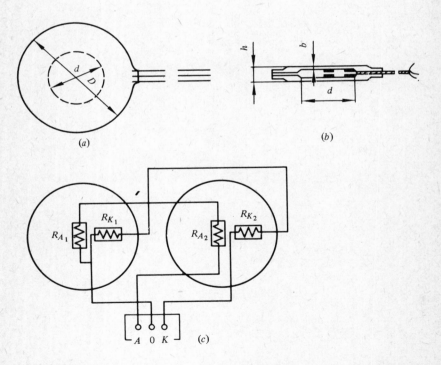

Fig. 90. Sketch of miniature dynamometer M.Ts.M.-2. (*a*) plan; (*b*) diametral cross-section; (*c*) system of connecting strain resistances.

Appendix 5

Fig. 91. General view of miniature dynamometer for centrifugal modelling, mark II (M.Ts.M.-2).

Bibliography

[1] A. M. Galperin. *Evaluation of the stability of open mine waste-heaps in conditions of the unstabilised state of water-saturated soil masses.* Moscow Institute of Radioelectronics and Mining Electromechanics, Moscow 1965.
[2] L. P. Zagorniko. 'The influence of the composition of soil mixtures on the stability of waste-heaps.' Coll. *Scientific notes*, issue 10. Kiev, Ukr.N.I.I. Projekt, 1963.
[3] L. P. Zagorniko. 'Determination of the conditions of stability of waste-heaps taking account of the technology and rate of their deposition.' Coll. *The stability of soil masses and the drying-out of open mines.* Kiev, Technika, 1968.
[4] V. V. Istomin. *The perfecting of methods of calculating the parameters of waste-heap formation in complex geotechnical conditions.* Moscow, Moscow Institute of Radioelectronics and Mining Electromechanics, 1966.
[5] R. P. Kaplunov and I. M. Panin. 'Concerning the problem of investigating the elements of mining operations by the centrifugal model testing method and the determination of the scale of time.' *Scientific works of the Moscow Mining Institute*, issue 8. Moscow, Ugletechizdat, 1950.
[6] Yu. N. Malushitsky. 'Procedure for laboratory investigations and calculations of the angles of waste-heap slopes.' In coll. *Scientific notes*, issue 5. Kiev, Ukr.N.I.I. Projekt, 1961.
[7] Yu. N. Malushitsky and A. P. Sakhno. 'The centrifugal model testing of waste-heaps.' Coll. reports *Problems of engineering geology*, issue 2. Leningrad, Geographical Society of the U.S.S.R., 1970.
[8] L. P. Markovich. *Investigations of the strength and stability of waste-heaps with reference to the technical mining factors at the pre-Carpathian sulphur mines.* Moscow, Moscow Institute of Radioelectronics and Mining Electromechanics, 1966.
[9] H. Matschak. *The stability of the slopes of faces and waste-heaps at open-cast lignite mines in the G.D.R.* Moscow, Central Scientific-Research Institute of Information and Technoeconomic Investigations in the Coal Industry, 1963.
[10] H. Matschak. Papers in coll. *Beiträge zur Geotechnik und Entwässerungstechnologie der Braunkohlentragebaue.* Freibergerforschungshafte. A. 407, 1967.
[11] V. V. Nalimov and N. A. Chernova. *Statistical methods of planning extreme experiments.* Moscow, Nauka, 1965.
[12] *Open workings at mining/chemical raw material sites* (Works of the Scientific Research Institute for Mining/Chemical Raw Material). Moscow, Gosgortechizdat, 1963.
[13] G. I. Pokrovsky and I. S. Fyodorov. *Centrifugal model testing for the solution of engineering problems.* Moscow, Grosstroiizdat, 1953.

Bibliography

[14] G. I. Pokrovsky and I. S. Fyodorov. *Centrifugal model testing in the construction industry*. Moscow, Gosstroiizdat, 1968.
[15] G. I. Pokrovsky and I. S. Fyodorov. *Centrifugal model testing in the mining industry*. Moscow, Niedra, 1969.
[16] A. A. Preobrazhensky *et al. Procedure for industrial experiments and for the construction of the simplest mathematical models in the investigation of mining machines and complexes*. Moscow, Moscow Mining Institute, 1969.
[17] RTM 44–62. *Procedure for the statistical processing of empirical data*. Moscow, Standards Publishing House, 1966.
[18] N. V. Smirnov, I. V. Dunin-Barkovsky. *A course on the theory of probability and mathematical statistics* (for technical applications). Moscow, Nauka, 1965.
[19] T. D. Ustinova. 'Investigations of the strength properties of soil mixtures in open-cast mines'. Coll. *Problems of engineering geology*, issue 2. Leningrad, Geographical Society of the U.S.S.R., 1970.
[20] E. Phillips. *Comptes Rendues*, vol. 68. Paris, 1869.
[21] G. M. Shakhunyants. *The earthen bed of railways*. Moscow, Franszheldorizdat, 1953.
[22] H. Schneckenberg. 'Der Einfluss der Bandtransportes auf die Festigkeit der transportierten Mischböden'. *Bergbautechnik,* no. 9, 1967.
[23] Ya. Yanko. *Mathematical–statistical tables*. Moscow, Gosstatizdat, 1961.

Index

abbreviated model of process, 194–5
'Absetz' waste-heap, 121
angle, critical, 3, 109
angle of collapse, critical, 139
angle of expression θ, 138–9, 156
angle of repose, 11, 62

Balakhov clayey loam, 100
basic parameters of slope of model, 184
bentonic clay, 69
bridge-formed waste-heap, 121
bulk forces of gravity, 1
bulk loading, 17
bulk weight, 2

calculation formula, 195–7
camoflet explosion, 157
capillarically wetted sand, 103, 105, 106, 107
centrifugal installation, 5
centrifugal machine, 4, 6, 7
centrifugal model testing installation, 4–11
centrifugal model testing method, 3, 116–24, 170, 175–81
  possible errors in, 177–9
centrifuge, acceleration of, 17
centrifuge model, 1
clayey loam, 69
  in fluid state, 62
  Novy Razdol, 61
clayey soil, profile of waste-heap model, 12
coefficient of compaction, 66
coefficient of compression, 152
coefficient of filtration, 152, 155
coefficient of linear consolidation, 59
coefficient of model, statistical analysis of, 193–4
coefficient of porosity, 152

coefficient of regression, 192
coefficient of reserve, 170, 172
coefficient of variation, 47, 55
collapse of waste-heap, 125, 167, 168
collapse point, 18
composition of soil mixture, 99–115
compression test, 147
consolidated slope, 119
consolidated waste-heap, 89
convex contour, bulging, 64, 79, 81
creeping tongue, 154
crevices, development of, 66
critical angle, 3, 109
  of collapse, 139
  of expression of pulp, 144
  steepness of, 156
critical height, 2, 81, 109, 111, 186
  of 'Absetz' waste-heap, 121
  of bridge-formed waste-heap, 121
critical moisture content of soil, 70, 71
critical stable profile of waste-heap, 136

deformation, 13
  of model, 60
  rupturing of non-sloping model, 48
degree of consolidation θ, 152, 154
degree of loading, 60
dense deposition of soil, 33, 117–18, 123
density of soil mass, 26, 29, 60, 84
density of waste-heap, 72–93
deposition of soil, average, 33
  dense, 33, 117–18
  loose, 33, 117
determination of properties of soil, 27
dimensionless variables, 192, 193
dispersion analysis method, 44, 187
dispersion ratio, 188
Dnieper sand, 29, 32, 73, 84, 99, 102, 110

Index

shear resistance in, 34
documentation, 27–9
drain, surface water, 157
'dry' waste-heap, 135–59
  measurement profile of, 142
  of non-saturable soil, 140
dyke, 157
  alluvially-deposited, 135
  protective, 135, 159
dynamometer, 198–9

embankment, artificially deposited, 3
empirical formulae, 70
excavator
  ESH-10/60, 140, 157, 160
  ESH-14/75, 66
  ESH-15/90, 160

factor experiment method, 93, 95, 189
factorial dispersion, 188
filtration coefficient, 152, 155
fissure, abrupt, 44
fissuring and crumpling of soil, 43
formation of waste-heap, 116–17
  rate of, 90
formula of normalisation, 191

granulometric composition of model, 141
  of waste-heap, 113

holding capacity
  of consolidated waste-heap, 89
  of unconsolidated waste-heap, 90
horizontal deformation of model, 22
hydraulic waste-heap, 134, 135, 138, 140, 141, 156, 158, 159
  Novobachat, 147
hydrophilic nature, 85
hydrostatic tub, 37, 104

inclined foundation, 72
indicator
  displacement (D.I.), 22, 24–5, 28, 121
  settlement (S.I.), 22, 24–5, 28, 121
inspection pit, 66, 67

Kaplunov–Panin method, 13, 14
Kerchi tertiary clay, 11, 29, 32, 41, 55, 68, 73, 74–5, 77, 81–2, 85–9, 91, 99, 102, 110, 113, 117, 184

Kiev, Ukr. N.I.I. Projekt Institute, 4
  centrifugal model testing installation, 11
Krasnogor mine, 135, 159, 160, 163, 166, 170, 172, 174

linear consolidation, 59–60, 61
  coefficient of, 59
lithologic composition of soil mass, 84–5
lithologic composition of waste-heap, 113
load-bearing capacity, 66
loading cycle, 18
loading of centrifuge carriages, 178
long tongue, 62
loose deposited model, deformation in time, 122
loose deposited waste-heap, 89–90
loose deposition of soil, 33, 117–18

Maslov shear device, 50
method, factor experiment, 93, 95
  Kaplunov–Panin, 13, 14
  paraffinisation, 47
method of deposition of waste-heap, stability on, 122–3
method of dispersion analysis, 44, 187
method of mathematical statistics, 69
method of waste-heap formation, 73
mixed waste-heap, 1, 11, 61, 115
model of unconsolidated waste-heap, 40–72
  shear in, 59
model testing method, centrifugal, 3, 116–24, 170, 175–81
model test procedure, 1–39
model waste-heap, 17
  after collapse, 10
  of strong skeletal soil, 128
  of variable compo-clay, 106
  on Novobachat hydraulic waste-heap, 147
modulus of elasticity, 13
moisture content of soil, 26, 27, 29, 33, 66, 72–93, 101, 102, 141, 150
  in non-sloping model, 57
multi-series model test, 67
multi-stage escarpment, 44
multi-step movement, 81

Index

natural angle of repose, 2
natural formation of waste-heap, 18
natural state, model of, 61–71
Neogene clay, 100
non-cohesive mass of soil, 167
non-sloping model, 11
    investigation of, 48–61
    settlement of, 52
normal loading, 53
Novobachat hydraulic waste-heap, model waste-heap on, 147
Novy Razdol alluvial sand, 61–2
Novy Razdol open cast mine, 29, 35, 62, 74
Novy Razdol quaternary soil, 11, 35, 61–2, 68, 75, 85, 94
    shear resistance in, 34
Novy Razdol soil, 89, 90, 91, 96, 113
    physico-mechanical properties of, 67
Novy Razdol sulphur deposit, 61, 73
Novy Razdol, tertiary soil, 11, 35, 61, 68, 75, 76, 80, 85, 86, 88, 91, 95, 128
    shear resistance in, 35
Novy Razdol waste-heap, 68, 94
    collapse of, 64, 65
    collapse of model of, 64, 65
    measurements on, 63

Okhotin's granulometric classification, 144
odometer, 147
overburden soil, 66, 140

paraffinisation method, 47
pellicular water, 103
penetrometer, 47
physico-mechanical indices, 177
physico-mechanical properties of soil, 29, 32, 139, 144
    of Krasnogor mine, 163
    of Novy Razdol soil, 67
    of waste-heap soil, 66
    recording of, 21
    using soil mechanics practice, 26
plasticity limit, 177
Pokrovsky's criterion, 1, 3
pore pressure, 72, 89
porosity coefficient, 152
profilometer, 21–2

Pronko machine, 36
protective dyke, 134
prototype, 61
prototype waste-heap, 68, 177
protrusion, 44, 80, 81, 131, 137
    lower terrace of, 46
pulp soil, 134–59
    physico-mechanical properties of, 139, 144

quaternary soil, 69
    Novy Razdol, 11, 35, 61–2, 68, 75, 85, 94

rate of formation of waste-heap, 72–93
recording of observations, 21–7
rupture, continuous curvilinear slickensided, 43
rupturing deformation, 48

sand, Dnieper, 29, 32, 73, 84, 99, 102, 110, 113
sandy loam, 44
scale factor, 47
scale of modelling, 16
secondary indices, 44
sequential approximation, 70
settlement, of non-sloping model, 52
settlement indicator (S.I.), 22, 24–5, 28, 121
shear, in model unconsolidated waste-heap, 59
    plastic, 48
shear plane, 53
shear resistance, 2, 27, 58, 79, 100, 108–9, 162
    in non-sloping model, 56
    of clayey, unconsolidated waste-heap, 61
    of Dnieper sand, 34
    of Novy Razdol quaternary soil, 34
    of soil mixtures, 101, 102,
    variation in, 105
shear resistance test, 53
skeletal soil, waste-heap of, 134
slickenside, 67, 79, 81
sliding tongue, 144, 157
sliding waste-heap, 62
slip, plan of surface, 28
    two parallel surfaces of, 43
slope displacement, 43

Index

sloping model, height of, 16
  investigation of, 40–8
soil, critical moisture content, 70, 71
  load bearing capacity, of, 66
soil composition, 109–13
soil crumpling and fissuring, 43
soil mass, density of, 26
soil material of model, 29–37
soil mixture of waste-heap, 99–104
stability, 100
  factors of unconsolidated waste-heap, 72
  indices of, 44, 169
  of unconsolidated waste-heap, 68, 72–93, 180, 181
  of waste-heap, 20, 84, 111, 116–24;
  on firm inclined base, 160, 99–115;
  on weak, inclined base, 115–33
stable profile, 144
stable waste-heap, 62
stroboscope, 7
surface of slip, 62
  continuous, 79, 81, 127
  roundly-continuous, 131
Svobodnensk sand and clay, 100

temporary stability, indices of, 170
tertiary clay
  Kerchi, 11, 29, 32, 41, 55, 68, 73, 74–5, 77, 81–2, 85–9, 91, 99, 102, 110, 113, 117, 184
  Tortonian, 62, 69, 73, 74, 78, 94, 113, 126
tertiary soil, Novy Razdol, 11, 35, 61, 68, 75, 76, 80, 85, 86, 88, 91, 95, 128
three-phase state, 14, 15, 55, 61
Tortonian tertiary clay, 62, 69, 73, 74, 78, 94, 113, 126
triangular cross-sectional waste-heap, 8
truncated cone waste-heap, 66
two-phase state, 14, 15, 55, 61

Ukr. N.I.I. Projekt Institute, 4, 11
unconsolidated waste-heap, 2, 3, 14, 89
  model of, 40–72
  stability factors in, 72
  stability of, 68, 72–93, 99–115

variation in shear resistance, 105
vertical deformation of model, 22

waste-heap, 'Absetz', 121
  bridge-formed, 121
  critical stable profile of, 136
  density of, 72–93
  'dry', 135–159; measurement profile of, 142; of non-saturable soil, 140
  formation of, 116–17
  formation rate of, 19, 72–93, 191
  granulometric composition of, 113
  hydraulic, 134, 135, 138, 140, 141, 156, 158, 159
  lithologic composition of, 113
  natural formation of, 18
  Novy Razdol: mixed, 115; quaternary, 85; tertiary, 85
  of 'ridged-jerky' character, 112
  of skeletal soils, 134
  sliding of, 62
  stability of, 20, 84
  stable, 62
  triangular cross-sectional, 8
  truncated cone, 66
  unconsolidated, 2, 3, 14, 89
waste-heap embankment, on weak foundation, 134–59
  stability of 175–81
waste-heap model, 17
  profile with clayey soil, 12
water-retaining capacity of sand and loam, weak foundation, waste-heap embankment on, 134–59